A2 [...]eek

Mathematics

Cath Brown and
[...]mingham
[...] Byrne

Where to find the information you need

Letts Educational
4 Grosvenor Place, London SW1X 7DL
School enquiries: 01539 564910
Parent & student enquiries: 01539 564913
E-mail: mail@lettsed.co.uk

Website: www.letts-educational.com

First published 2001
10 9 8 7 6 5

British Library Cataloguing in Publication Data
A CIP record for this book is available from the British Library.

ISBN 978-1-84315-816-5

Cover design by Purple, London

Prepared by *specialist* publishing services, Milton Keynes

Printed in Dubai

Test your knowledge

1
a) i) Expand $(2 + 3x)^6$ up to the term in x^3
ii) Hence find an approximation to 2.03^6
b) i) Expand $(2 - 3x)^{\frac{1}{2}}$ up to the term in x^3, giving your answer in terms of the simplest possible surds and fractions.
ii) State the values of x for which your expansion is valid.

2
a) Find the remainder when $2x^3 - 2x^2 + 7x - 5$ is divided by $(x + 2)$
b) The remainder when $x^4 - 4x^3 + Ax - 5$ is divided by $(x - 1)$ is 1. Find the value of the constant A.

3 Express in partial fractions:

a) $\dfrac{5x + 37}{(x - 1)(x + 6)}$ b) $\dfrac{3x + 14}{(x^2 + 6)(x - 2)}$

c) $\dfrac{5x^2 + 21x + 13}{(x + 3)^2(x - 2)}$ d) $\dfrac{6x^2 + 2x - 3}{(x - 2)(x + 3)}$

4
a) Express as a single logarithm: $\log 2 + \log 6 - 2\log 3$
b) Express in terms of simple logarithms: $\log(a^2b^3 \div c)$
c) Make y the subject of the following:
i) $\ln y + 2\ln x = 2\ln 4$
ii) $2e^{y+1} = x + 1$
d) Solve the equation: $8^x = 3$ giving your answer correct to 2DP.

Answers

1 a) i) $64 + 576x + 2160x^2 + 4320x^3$ **ii)** 69.98032

b) i) $\sqrt{2}(1 - 3x/4 - 9x^2/32 - 27x^3/128)$ **ii)** $-\frac{2}{3} \leq x < \frac{2}{3}$

2 a) -43 **b)** $A = 9$

3 a) $\frac{6}{(x-1)} - \frac{1}{(x+6)}$ **b)** $\frac{(-2x-1)}{(x^2+6)} + \frac{2}{(x-2)}$

c) $\frac{1}{(x+3)^2} + \frac{2}{(x+3)} + \frac{3}{(x-2)}$ **d)** $6 + \frac{5}{(x-2)} - \frac{6}{(x+3)}$

4 a) $\log(\frac{4}{3})$ **b)** $2\log a + 3\log b - \log c$

c) i) $y = \frac{16}{x^2}$ **ii)** $y = \ln[\frac{1}{2}(x+1)] - 1$ **d)** $x = 0.53$

If you got them all right, skip to page 12

Improve your knowledge

1 The Binomial Theorem

This theorem tells you how to expand a bracket of the form $(1 + x)^n$. It says:

$$(1 + x)^n = 1 + nx + \frac{n(n-1)}{2!}x^2 + \frac{n(n-1)(n-2)}{3!}x^3 + \frac{n(n-1)(n-2)(n-3)}{4!}x^4 + \ldots$$

Example 1

Find the expansion of $(1 + x)^{-3}$ up to the term in x^3

Solution

$$(1 + x)^{-3} = 1 + -3x + \frac{-3(-3-1)}{2!}x^2 + \frac{-3(-3-1)(-3-2)}{3!}x^3$$

$$= 1 - 3x + 6x^2 - 10x^3$$

$n!$ means $n \times (n-1) \times (n-2) \times \ldots \times 2 \times 1$. So $5!$ means $5 \times 4 \times 3 \times 2 \times 1 = 120$

Most of the time, you are not just asked to expand $(1 + x)^n$ – instead you are given something like $1 - 2x$ or $2 + 3x$ inside the bracket.

Expanding a bracket with a 1 at the front (like $(1 - 2x)^5$)

With these, you use the formula in exactly the same way, but instead of x, x^2, x^3 etc., you'd have $(-2x)$, $(-2x)^2$, $(-2x)^3$, etc. (so if it was $(1 + \frac{1}{2}x)^{-2}$, you'd have $\frac{1}{2}x$ in each bracket).

Example 2

Expand $(1 - 2x)^5$ up to the term in x^3

Solution

$$(1 - 2x)^5 = 1 + 5(-2x) + \frac{5(5-1)}{2!}(-2x)^2 + \frac{5(5-1)(5-2)}{3!}(-2x)^3$$

$$= 1 - 10x + 40x^2 - 80x^3$$

You MUST put the brackets in. You are almost certain to get it wrong otherwise

Expanding a bracket that doesn't have a 1 at the front (like $(2 - 3x)^5$)

With these, before you start, you have to take out a factor so that the bracket you expand does start with a 1. Example 3 illustrates how to do this.

Example 3

Expand $(2 + 3x)^5$ up to the term in x^3

Solution

$$(2 + 3x) = 2\left(1 + \frac{3x}{2}\right)$$

$$(2 + 3x)^5 = 2^5\left(1 + \frac{3x}{2}\right)^5$$

$$= 2^5\left[1 + 5\left(\frac{3x}{2}\right) + \frac{5(5-1)}{2!}\left(\frac{3x}{2}\right)^2 + \frac{5(5-1)(5-2)}{3!}\left(\frac{3x}{2}\right)^3\right]$$

$$= 32\left[1 + \frac{15x}{2} + \frac{45x^2}{2} + \frac{135x^3}{4}\right]$$

$$= 32 + 240x + 720x^2 + 1080x^3$$

Don't forget that the number in front must be to the power as well!

Do NOT try to be 'clever' by combining powers. Just work things out the long way!

Differences depending on whether the power is a whole, positive number

If the power is a whole, positive number

- the expansion is valid for any value of x you put into it
- the expansion will eventually stop, e.g. $(1 + x)^3 = 1 + 3x + 3x^2 + x^3$

If the power is a whole, positive number, the question may just ask you to expand the bracket, without saying how many terms to use. If this happens, you just carry on until you get 0.

If the power is a fraction or a negative number

- the expansion goes on forever, e.g. $(1 - x)^{-1} = 1 + x + x^2 + x^3 + x^4 + \ldots$
- the expansion is only valid for certain values of x

In this sort of example, you may be asked to give the values of x for which the expansion is valid. To work this out:

1) take out a factor so the bracket has a 1 at the front
2) if there's a + sign: write down $-1 <$ other thing in bracket ≤ 1
   if there's a − sign: write down $-1 \leq$ other thing in bracket < 1
3) multiply or divide as necessary to find the values of x.

Example 4

State the values of x for which the binomial expansions of each of the following are valid:

a) $(3 + 4x)^{10}$ b) $(1 - 2x)^{\frac{1}{2}}$ c) $(4 + 3x)^{-2}$

Solution

a) All values of x, as the power is a positive whole number

b) $-1 \leq 2x < 1 \Rightarrow -\frac{1}{2} \leq x < \frac{1}{2}$

c) $4 + 3x = 4(1 + \frac{3}{4}x)$

So $-1 < \frac{3}{4}x \leq 1 \Rightarrow -\frac{4}{3} < x \leq \frac{4}{3}$

Some common types of question

These are 'standard' types of example. You need to learn how to do them.

Example 5

a) Find the first four terms in the expansion of $(1 + 2x)^{10}$

b) Hence obtain an approximation to 1.02^{10}

Solution

a) $$(1 + 2x)^{10} = 1 + 10(2x) + \frac{10(10-1)}{2!}(2x)^2 + \frac{10(10-1)(10-2)}{3!}(2x)^3$$

$$= 1 + 20x + 180x^2 + 960x^3$$

b) We must use the first part. Let $(1 + 2x)^{10} = 1.02^{10}$. So $1 + 2x = 1.02$. So $2x = 0.02$. So $x = 0.01$.

Put $x = 0.01$ in expansion:
$(1.02)^{10} \approx 1 + 20\,(0.01) + 180(0.01)^2 + 960(0.01)^3 = 1.21896$

Example 6

The 'coefficient' of x is just all the things in front of x. It DOESN'T include the x itself

In the expansion of $(1 + ax)^n$, the coefficients of x and x^2 are -6 and 27 respectively. Find the values of a and n.

Solution

Since it refers to an expansion, we must expand what we have:

$(1 + ax)^n = 1 + nax + \frac{1}{2}n(n-1)(ax)^2$

Now we use what we are given:

Coefficient of $x = na = -6$

Coefficient of $x^2 = \frac{1}{2}n(n-1)a^2 = 27$

Now we must solve these equations:

We have $a = -\frac{6}{n}$

So $\dfrac{36n(n-1)}{2n^2} = 27$

Cancelling and multiplying up: $18(n-1) = 27n \Rightarrow n = -2, a = 3$

2 The Remainder Theorem

The remainder theorem is about what remainder you get when you divide a polynomial (i.e. something with x^3, etc., in) by a linear factor – that's something like $(x-3)$ or $(2x-1)$. It says:

When you divide a polynomial $f(x)$ by $(bx-a)$, the remainder is $f(\frac{a}{b})$
So if you were dividing by $(x-3)$, you'd put $x = 3$; if by $(x+1)$, you'd put $x = -1$ and if by $(2x-1)$, you'd put $x = \frac{1}{2}$
Examples 7 and 8 show how this is commonly used.

Example 7

a) Find the remainder when $x^4 - 3x^3 + 2x^2 - 5x + 6$ is divided by $(x+2)$
b) The remainder when $2x^3 - 3x^2 + 8x - A$ is divided by $(2x+1)$ is -3. Find A

Solution

a) To find the remainder, we put in $x = -2$:
$(-2)^4 - 3(-2)^3 + 2(-2)^2 - 5(-2) + 6 = 64$

b) We put $x = -\frac{1}{2}$, and the answer must come to -3:
$2(-\frac{1}{2})^3 - 3(-\frac{1}{2})^2 + 8(-\frac{1}{2}) - A = 3 \Rightarrow -\frac{1}{4} - \frac{3}{4} - 4 - A = -3 \Rightarrow A = -2$

Example 8

$f(x) \equiv x^3 + Ax^2 + Bx - 5$

When $f(x)$ is divided by $(x-1)$, the remainder is -6. When $f(x)$ is divided by $(x+1)$, the remainder is 4. Find A and B

Solution

Put $x = 1$, get -6:
$(1)^3 + A(1)^2 + B(1) - 5 = -6 \Rightarrow A + B - 4 = -6 \Rightarrow A + B = -2$ **(1)**
Put $x = -1$, get 4:
$(-1)^3 + A(-1)^2 + B(-1) - 5 = 4 \Rightarrow A - B - 6 = 4 \Rightarrow A - B = 10$ **(2)**

Adding (1) and (2): $2A = 8 \Rightarrow A = 4$. Substituting back $\Rightarrow B = -6$

Partial Fractions

Normal Partial Fractions

Partial fractions involve splitting up something like $\frac{3}{(x+1)(x-2)}$ into $\frac{1}{(x-2)} - \frac{1}{(x+1)}$

There are three different types that you must know; the types are split according to what the denominator (bottom) of the fraction is like.

Type I: Linear factors like $\frac{3}{(x+1)(x-2)}$ Use $\frac{A}{(x-2)} + \frac{B}{(x+1)}$

Type II: Squared inside bracket like $\frac{3}{(x^2+2)(x-3)}$ Use $\frac{Ax+B}{(x^2+2)} + \frac{C}{(x-3)}$

In type II, the (Ax + B) MUST be over the denominator with the x^2 in it

Type III: Squared outside bracket like $\frac{3}{(x-1)^2(x+4)}$

Use $\frac{A}{(x-1)^2} + \frac{B}{(x-1)} + \frac{C}{(x+4)}$

In each case, you put the partial fractions over the denominator you started with (this is important), and substitute in values of x to find the constants A, B and C. Example 9 illustrates the method.

Example 9

Express as partial fractions

a) $\frac{x-5}{(x-2)(x-3)}$ b) $\frac{2x^2+x+11}{(x^2+3)(x+1)}$ c) $\frac{x^2-9x+17}{(x-2)^2(x+1)}$

Solution

a) $$\frac{x-5}{(x-2)(x-3)} \equiv \frac{A}{(x-2)} + \frac{B}{(x-3)} \equiv \frac{A(x-3)+B(x-2)}{(x-2)(x-3)}$$

So $x - 5 \equiv A(x-3) + B(x-2)$

Now we choose values of x to substitute in. Wherever possible, we try to make one of the brackets zero.

$x - 5 \equiv A(x-3) + B(x-2)$

$x = 3$ $3 - 5 = A(3-3) + B(3-2) \Rightarrow -2 = A(0) + B \Rightarrow B = -2$

$x = 2$ $2 - 5 = A(2-3) + B(2-2) \Rightarrow -3 = A(-1) + B(0) \Rightarrow A = 3$

b) $\dfrac{2x^2 + x + 11}{(x^2 + 3)(x + 1)} \equiv \dfrac{(Ax + B)}{(x^2 + 3)} + \dfrac{C}{(x + 1)} \equiv \dfrac{(Ax + B)(x + 1) + C(x^2 + 3)}{(x^2 + 3)(x + 1)}$

Make sure you put the brackets around the (Ax + B) – or you'll make a mistake!

So $2x^2 + x + 11 \equiv (Ax + B)(x + 1) + C(x^2 + 3)$

$x = -1$: $2(-1)^2 + (-1) + 11 = (A(-1) + B)(-1 + 1) + C((-1)^2 + 3)$
$\Rightarrow 12 = (-A + B)(0) + C(4) \Rightarrow 12 = 4C \Rightarrow C = 3$

We cannot make the other bracket zero, so choose other easy values of x. Always start with $x = 0$:

$x = 0 \quad 2(0)^2 + (0) + 11 = (A(0) + B)(0 + 1) + C((0)^2 + 3) \Rightarrow 11 = B(1) + C(3)$

But $C = 3 \Rightarrow 11 = B + 9 \Rightarrow B = 2$

$x = 1 \quad 2(1)^2 + (1) + 11 = (A(1) + B)(1 + 1) + C((1)^2 + 3)$
$\Rightarrow 14 = (A + B)(2) + C(4)$

But $C = 3$, $B = 2 \Rightarrow 14 = 2A + 4 + 12 \Rightarrow A = -1$

c) $\dfrac{x^2 - 9x + 17}{(x - 2)^2(x + 1)} \equiv \dfrac{A}{(x - 2)^2} + \dfrac{B}{(x - 2)} + \dfrac{C}{x + 1}$

With this type we must take special care putting them over a common denominator. The denominator we use is $(x - 2)^2(x + 1)$. We divide this by each of the other denominators in turn to work out what to multiply A, B and C by:

For A: $(x - 2)^2(x + 1) \div (x - 2)^2 = (x + 1)$, so multiply A by $(x + 1)$
For B: $(x - 2)^2(x + 1) \div (x - 2) = (x + 1)(x - 2)$
For C: $(x - 2)^2(x + 1) \div (x + 1) = (x - 2)^2$

So we get

$$\frac{x^2 - 9x + 17}{(x - 2)^2(x + 1)} \equiv \frac{A}{(x - 2)^2} + \frac{B}{(x - 2)} + \frac{C}{x + 1} \equiv \frac{A(x + 1) + B(x + 1)(x - 2) + C(x - 2)^2}{(x - 2)^2(x + 1)}$$

so $x^2 - 9x + 17 \equiv A(x + 1) + B(x + 1)(x - 2) + C(x - 2)^2$

Put $x = -1$ $\quad (-1)^2 - 9(-1) + 17 \equiv A(-1 + 1) + B(-1 + 1)(-1 - 2) + C(-1 - 2)^2$
$\Rightarrow 27 = A(0) + B(0) + 9C \Rightarrow C = 3$

Put $x = 2$: $\quad (2)^2 - 9(2) + 17 \equiv A(2 + 1) + B(2 + 1)(2 - 2) + C(2 - 2)^2$
$\Rightarrow 3 = A(3) + B(0) + C(0) \Rightarrow A = 1$

Put $x = 0$: $\quad (0)^2 - 9(0) + 17 \equiv A(0 + 1) + B(0 + 1)(0 - 2) + C(0 - 2)^2$
$\Rightarrow 17 = A(1) + B(-2) + C(4)$

But $A = 1$, $C = 3 \Rightarrow 17 = 1 - 2B + 12 \Rightarrow B = -2$

Top-Heavy Partial Fractions

A partial fraction is top-heavy if the highest power of x on the top is the same as the highest power of x on the bottom (if all the brackets were multiplied out). If you get one like this, you have to start with $\frac{A}{1}$, then continue with the normal partial fractions. You do exactly the same as before then – common denominator, substitute in, etc.

Always check whether it is top-heavy before you start!

Example 10

Express in partial fractions $\dfrac{2x^2 - 6x + 8}{(x-1)(x+1)}$

Solution

This is top-heavy since both top and bottom would have x^2 terms if it was multiplied out.

So set $\dfrac{2x^2 - 6x + 8}{(x-1)(x+1)} \equiv \dfrac{A}{1} + \dfrac{B}{(x-1)} + \dfrac{C}{x+1}$

$$\equiv \frac{A(x-1)(x+1) + B(x+1) + C(x-1)}{(x-1)(x+1)}$$

$$\Rightarrow 2x^2 - 6x + 8 \equiv A(x-1)(x+1) + B(x+1) + C(x-1)$$

Put $x = 1$: $2(1)^2 - 6(1) + 8 = A(1-1)(1+1) + B(1+1) + C(1-1)$
$\Rightarrow 4 = A(0) + B(2) + C(0) \Rightarrow B = 2$

Put $x = -1$: $2(-1)^2 - 6(-1) + 8 = A(-1-1)(-1+1) + B(-1+1) + C(-1-1)$
$\Rightarrow 16 = A(0) + B(0) + C(-2) \Rightarrow C = -8$

Put $x = 0$: $2(0)^2 - 6(0) + 8 = A(0-1)(0+1) + B(0+1) + C(0-1)$
$\Rightarrow 8 = A(-1) + B(1) + C(-1)$

But $B = 2$, $C = -8 \Rightarrow 8 = -A + 2 + 8 \Rightarrow A = 2$

4 Logarithms and Exponentials

You need to know the laws of logs, which are:

$\log(ab) = \log a + \log b$ $\quad$ $\log(\frac{a}{b}) = \log a - \log b$ $\quad$ $\log(a^n) = n\log a$

In other words, if you are multiplying the numbers, add the logs; if you are dividing, subtract them, and you can 'bring down' powers.

In addition to the laws, you need to know that:

$\log 1 = 0$.
$\log_a a = 1$, whatever a (the base of the logarithm) is
You cannot have logs of negative numbers, or zero

You can only manipulate logs according to these rules. For example, you can do nothing with $\log(a + b)$.

You will mainly deal with $\ln x$, which is the 'opposite' (or more formally, the inverse function of) e^x. This means that $e^{\ln x} = \ln e^x = x$ – so e and ln 'cancel'.

NB This cancelling only works when there is nothing in between the e and the ln – you cannot 'cancel' $e^{-\ln x}$ or $\ln(2e^x)$ directly

You can use this to get rid of ln or e in an equation – but you have to be very careful how you do it!

Example 11

a) Given that $\ln y = 2\ln x + \ln 5$, express y in terms of x
b) Given that $e^{3y+4} - 5 = 2x$, express y in terms of x

Solution

a) Before trying to get rid of the lns, we need to get both sides in the form ln(something).

   We know $2\ln x = \ln x^2$ (laws of logs).
   So $\ln y = \ln(x^2) + \ln 5 = \ln(5x^2)$ (laws of logs)

   Now we can take exponentials: $e^{\ln y} = e^{\ln(x^2)} \Rightarrow y = 5x^2$

b) We will need to take logs to get rid of the e, but first we must make sure the $e^{something}$ is on its own:

   $e^{3y+4} - 5 = 2x \quad \Rightarrow e^{3y+4} = 2x + 5$

   Taking logs: $\quad \Rightarrow \ln(e^{3y+4}) = \ln(2x + 5)$

   So $3y + 4 = \ln(2x + 5) \Rightarrow y = \dfrac{\ln(2x + 5) - 4}{3}$

NB Brackets are vital – it is NOT $\ln 2x + \ln 5$

   Logs can also be used to solve equations with powers in them:

Example 12

Solve the equation $3^x = 4^{\frac{1}{x}}$; $x > 0$ giving your answer correct to 2 decimal places.

Solution

To 'bring powers down', take logs of each side:

$\ln(3^x) = \ln(4^{\frac{1}{x}}) \Rightarrow x\ln 3 = \frac{1}{x}\ln 4$

Rearranging: $\quad x^2 = \dfrac{\ln 4}{\ln 3} \Rightarrow x = 1.12$ (2DP)

Algebra

45 minutes

Use your knowledge

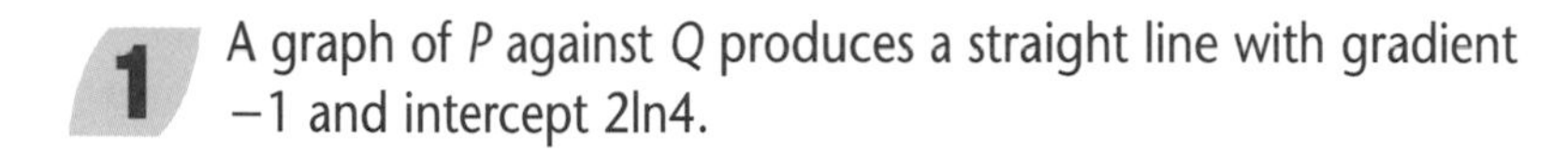

1 A graph of P against Q produces a straight line with gradient -1 and intercept $2\ln 4$.

a) Write down an equation relating P and Q

If you are stuck on this, revise AS coordinate geometry

b) Given that $P = \ln y$ and $Q = \ln x$, find y in terms of x, giving your answer in a form not involving logarithms.

Substitute into your equation from a)
Get each side in the form ln(something)

2 a) Expand $(1 - 4x)^7$ up to the term in x^3
Hence

b) i) Obtain an approximation to 0.96^7

You need $(1 - 4x)^7 = 0.96^7$ – find x

ii) Calculate the percentage error in using this approximation.

$\%\ \text{error} = \dfrac{\text{approx} - \text{accurate}}{\text{accurate}} \times 100\%$

c) Show that $\left(1 + \dfrac{4}{\sqrt{3}}\right)^7 + \left(1 - \dfrac{4}{\sqrt{3}}\right)^7 \approx 226$

Write down the expansion of $(1 + 4x)^7$
Find $(1 + 4x)^7 + (1 - 4x)^7$
Substitute in for x

3 $f(x) \equiv x^3 + Ax^2 + Bx + 6$ is divisible by $(x + 1)$, and has remainder 60 when divided by $(x - 2)$

a) Find the values of A and B

Substitute in $x = -1$ and $x = 2$ to give two simultaneous equations
You may need to revise the Factor Theorem from AS

b) Factorise $f(x)$ fully.

You already know $(x + 1)$ is a factor!

4 a) Express $\dfrac{2x - 5}{x - 4}$ in the form $A + \dfrac{B}{x - 4}$, where A and B are constants to be determined.

Do this the same way as top-heavy partial fractions

b) Hence or otherwise, obtain the expansion of $\dfrac{2x - 5}{x - 4}$ in ascending powers of x up to the term in x^2. State the range of values for which your expansion is valid.

Use part a)
What power is the $(x - 4)$ to?
Rearrange the $x - 4$ to be $-4 + x$

Test your knowledge

1 Find the solutions to the following equation: $\sin(x - \frac{\pi}{9}) = -0.5$; $-2\pi < x < 2\pi$

2 Find the solution to the following equations:

a) $\sec 3x = 2$; $-180° < x < 180°$

b) $\mathrm{cosec}^2 2x = 2\cot 2x$; $-\pi \leq x \leq \pi$

c) $\mathrm{cosec}\,x = 2\sin x$; $-\pi \leq x \leq \pi$

3 Find the solution to the following equations:

a) $\cos 2x = \cos x$; $-\pi \leq x \leq \pi$

b) $\tan 2x = 3\tan x$; $0° \leq x \leq 360°$

c) $\sin(x - 30) = 2\cos x$; $0° \leq x \leq 360°$

4 Prove the following identities:

a) $\cos 3x \equiv 4\cos^3 x - 3\cos x$

b) $\sec^2 x\,\mathrm{cosec}^2 x \equiv \sec^2 x + \mathrm{cosec}^2 x$

c) $\sin(x + 30) + \sin(x - 30) \equiv \sqrt{3}\sin x$

5 Express $\sqrt{3}\sin x - \cos x$ in the form $R\sin(x - \alpha)$, where R is a positive constant and α is an acute angle.

Answers

1 $x = \frac{35\pi}{18}, -\frac{\pi}{18}, \frac{23\pi}{18}, -\frac{13\pi}{18}$

2 a) $x = -140°, -100°, -20°, 20°, 100°, 140°$

b) $x = \frac{\pi}{8}, \frac{5\pi}{8}, -\frac{7\pi}{8}, -\frac{3\pi}{8}$ **c)** $x = \pm\frac{\pi}{4}, \pm\frac{3\pi}{4}$

3 a) $x = \frac{2\pi}{3}, -\frac{2\pi}{3}, 0$ **b)** $x = 0, 180°, 30°, 210°, 150°, 330°, 360°$

c) $x = 70.9°, 250.9°$

4 a) $\mathrm{LHS} \equiv \cos 3x \equiv \cos(2x + x) \equiv \cos 2x\cos x - \sin 2x\sin x$

$\equiv (2\cos^2 x - 1)\cos x - 2\sin x\cos x\sin x \equiv 2\cos^3 x - \cos x - 2\sin^2 x\cos x$

Substituting $\sin^2 x = 1 - \cos^2 x$: $\mathrm{LHS} \equiv 2\cos^3 x - \cos x - 2(1 - \cos^2 x)\cos x$

$\equiv 2\cos^3 x - \cos x - 2\cos x + 2\cos^3 x = 4\cos^3 x - 3\cos x = \mathrm{RHS}$

b) $\mathrm{LHS} \equiv \frac{1}{\cos^2 x}\frac{1}{\sin^2 x} \equiv \frac{1}{\sin^2 x\cos^2 x}$

$\mathrm{RHS} \equiv \frac{1}{\cos^2 x} + \frac{1}{\sin^2 x} \equiv \frac{\sin^2 x + \cos^2 x}{\sin^2 x\cos^2 x} \equiv \frac{1}{\cos^2 x\sin^2 x} \equiv \mathrm{LHS}$

c) $\mathrm{LHS} \equiv \sin(x + 30) + \sin(x - 30) \equiv \sin x\cos 30 + \cos x\sin 30 + \sin x\cos 30$

$- \cos x\sin 30 \equiv 2\sin x\cos 30 \equiv 2\sin x\frac{\sqrt{3}}{2} \equiv \sqrt{3}\sin x$

5 $R = 2$; $\alpha = 30°$

If you got them all right, skip to page 20

Improve your knowledge

You need to revise how to solve all the kinds of trigonometric equations you met in AS.

1 Working in Radians

If the question is set in radians, you MUST give your answer in radians. In particular, questions involving trig functions in calculus or numerical methods are always in radians.

If a question requires you to work in radians, you must generally give your answer in terms of π (unless it says something like 'give your answer to 2 decimal places').

If you don't like working in terms of radians throughout a question, you can work in degrees, but you must remember to convert back at the end.

Remember – convert from degrees to radians by multiplying by $\frac{\pi}{180}$

Remember:– to convert degrees to radians, multiply by $\frac{\pi}{180}$

to convert radians to degrees, multiply by $\frac{180}{\pi}$

2 Reciprocal Trig Ratios

These are $\sec x = \dfrac{1}{\cos x}$; $\cot x = \dfrac{1}{\tan x}$ and $\operatorname{cosec} x = \dfrac{1}{\sin x}$

(To remember what sec, cosec and cot are, look at the third letter!)

Equations involving these will either involve using the above definitions, or the squared ratio identities. These may be in your formula book, and are:

$\tan^2 A + 1 = \sec^2 A$; $\quad 1 + \cot^2 A = \operatorname{cosec}^2 A$

You have already met the third (and most important) one of these: $\sin^2 A + \cos^2 A = 1$

You use the squared ratio identities if:

i) The equation contains sin and cos, or sec and tan, or cosec and cot (it must be of the same angle)

AND ii) one (or more) of them is squared, e.g. you have sec and $\tan^2$

Otherwise, you should try putting everything in terms of sin and cos, and attempting to simplify.

Example 1

Solve the equations

a) $\sec x = 1 + 2\tan^2 x \quad -180° < x < 180°$ b) $2\cos 2x = \sec 2x \quad 0 < x < \pi$

Solution

a) Since equation contains sec and tan, and one of them is squared, use $\tan^2 A + 1 = \sec^2 A$:

$\sec x = 1 + 2\tan^2 x = 1 + 2(\sec^2 x - 1) = 1 + 2\sec^2 x - 2 = 2\sec^2 x - 1$

So $\sec x = 2\sec^2 x - 1 \Rightarrow 2\sec^2 x - \sec x - 1 = 0 \Rightarrow (2\sec x + 1)(\sec x - 1) = 0$

$\Rightarrow \sec x = -\frac{1}{2}$ or 1

To find a solution, we need to change to 'normal' trig functions:

$\frac{1}{\cos x} = \sec x = -\frac{1}{2}$ or $1 \Rightarrow \cos x = -2$ or 1.

$\cos x = -2$ gives no solutions; $\cos x = 1$ gives $x = 0$

b) Can't use the squared ratio identities $\Rightarrow$ try putting everything in terms of sin and cos

(NB: It doesn't matter that it's $2x$, not x – it would only be a problem if we had, say, $\cos x$ and $\sec 2x$)

$$2\cos 2x = \sec 2x \Rightarrow 2\cos 2x = \frac{1}{\cos 2x} \Rightarrow 2\cos 2x \times \cos 2x = 1$$

$$\Rightarrow \cos^2 2x = \frac{1}{2}$$

$$\Rightarrow \cos 2x = \pm\sqrt{\tfrac{1}{2}}$$

Always remember the $\pm$ when you take a square root – missing it out loses lots of marks!

So solutions are $2x = 45°, 135°, 225°, 315° \Rightarrow x = 22\frac{1}{2}°, 77\frac{1}{2}°, 112\frac{1}{2}°, 157\frac{1}{2}°$

Converting to radians: $\frac{\pi}{8}, \frac{3\pi}{8}, \frac{5\pi}{8}, \frac{7\pi}{8}$

3 Addition Formulae

The addition formulae are in your formula book. They are:

$\sin(A \pm B) = \sin A\cos B \pm \cos A\sin B$

$\cos(A \pm B) = \cos A\cos B \mp \sin A\sin B$

$$\tan(A \pm B) = \frac{\tan A \pm \tan B}{1 \mp \tan A\tan B}$$

NB: The + and − signs are the other way up in some formulae for a reason!

$\cos(A + B) \equiv \cos A\cos B - \sin A \sin B$

From these, you can derive (by putting $B = A$) the double-angle identities:

$$\sin 2A \equiv 2\sin A\cos A; \qquad \cos 2A \equiv \cos^2 A - \sin^2 A; \qquad \tan 2A \equiv \frac{2\tan A}{1 - \tan^2 A}$$
$$\equiv 2\cos^2 A - 1$$
$$\equiv 1 - 2\sin^2 A$$

A can be any angle – 30°, x, 3x…

If these are not in your formula book, you need to learn them.

You also need to be able to derive them, as shown above.

You know to use the addition formulae or the double-angle formulae in an equation when you see something like $\sin(x + 30)$, $\sin 2x$ or $\sin 3x$ (you split this one as $\sin(2x + x)$)

NB: You only use these formulae if you have, say, a mixture of $2x$ (or $(x + 30)$, etc.) and other things. If the equation involves only one thing, e.g. $\sin 2x = 3\cos 2x$ – you don't use them.

Example 2

Find the solution to the following equations for the specified values of x:

a) $5\sin 2x - 4\sin x = 0 \qquad -180° \leq x \leq 180°$
b) $\sin(x + 60) = 2\cos(x - 45) \qquad 0° < x < 360°$

Solution

a) Since we have a mixture of x and $2x$, use double-angle formulae:
$5\sin 2x - 4\sin x = 0 \Rightarrow 5(2\sin x\cos x) - 4\sin x = 0$
$\Rightarrow 10\sin x\cos x - 4\sin x - 0 \Rightarrow \sin x(10\cos x - 4) = 0$
Hence $\sin x = 0$ or $10\cos x - 4 = 0 \Rightarrow \sin x = 0$ or $\cos x = \frac{4}{10}$
$\sin x = 0 \Rightarrow x = 0°, 180°, -180°; \qquad \cos x = \frac{4}{10} \Rightarrow x = 66.4°, -66.4°$

Never cancel the sinx in a case like this – you lose a solution

b) Since we do not have the same thing on both sides, we need to expand both:
$\sin(x + 60) \equiv \sin x\cos 60 + \cos x\sin 60 \equiv \frac{1}{2}\sin x + \frac{\sqrt{3}}{2}\cos x$
$\cos(x - 45) \equiv \cos x\cos 45 + \sin x\sin 45 \equiv \frac{1}{\sqrt{2}}\cos x + \frac{1}{\sqrt{2}}\sin x$
So $\frac{1}{2}\sin x + \frac{\sqrt{3}}{2}\cos x \equiv \frac{2}{\sqrt{2}}\cos x + \frac{2}{\sqrt{2}}\sin x$
Collecting terms: $\sin x(\frac{1}{2} - \frac{2}{\sqrt{2}}) = \cos x(\frac{2}{\sqrt{2}} - \frac{\sqrt{3}}{2})$
Dividing:

$$\tan x = \frac{\sin x}{\cos x} = \frac{\left(\frac{2}{\sqrt{2}} - \frac{\sqrt{3}}{2}\right)}{\left(\frac{1}{2} - \frac{2}{\sqrt{2}}\right)} = -0.5996$$

$\tan^{-1}(-0.5996) = -30.9° \Rightarrow x = 149.1°, 329.1°$

You need to know and use:
$\cos 30 = \sin 60 = \frac{\sqrt{3}}{2}$ (= 0.866…)
$\sin 45 = \cos 45 = \frac{1}{\sqrt{2}}$ (= 0.707)

Identities

Identities are things you have to prove; you don't have to find values of x. Identities are written with a ≡ sign (not a = sign).

When you are given an identity to prove, there isn't any absolutely fail-safe strategy, but the following is always a good idea:

- Get everything in terms of sin and cos
- If you have (say) $2x$ on one side and x on the other, use addition/double-angle formulae
- Put any fractions over a common denominator
- Try using $\sin^2 + \cos^2 = 1$
- Use the answer to help you; if it wants just cos, try to find a way to get rid of everything else!

If you're stuck on an equation, try these tips too!

Because you are trying to prove something, you shouldn't start out assuming it; it is a good idea to use 'LHS ≡' and 'RHS ≡' (LHS = left-hand side). However, you can work on both sides and get them to meet in the middle – just don't work on both at once!

Example 3

Prove the following identities:

a) $\tan x + \cot x \equiv 2\operatorname{cosec} 2x$

b) $\sin 3x \equiv 3\sin x - 4\sin^3 x$

Solution

a) Putting everything in terms of sin and cos:

$$LHS \equiv \frac{\sin x}{\cos x} + \frac{\cos x}{\sin x} \qquad RHS \equiv \frac{2}{\sin 2x}$$

Using double-angle formulae:

$$RHS \equiv \frac{2}{2\sin x\cos x} \equiv \frac{1}{\sin x\cos x}$$

This seems as far as we're likely to get with the RHS, so work on LHS:

$$LHS \equiv \frac{\sin x}{\cos x} + \frac{\cos x}{\sin x} \equiv \frac{\sin^2 x + \cos^2 x}{\sin x\cos x} = \frac{1}{\sin x\cos x} \equiv RHS$$ (put over a common denominator and using $\sin^2 A + \cos^2 A = 1$)

b) Everything is already in terms of sin and cos. Since we have $\sin 3x$ on one side and $\sin x$ on the other, need to use addition formulae:

$$\text{LHS} \equiv \sin 3x \equiv \sin(2x + x) \equiv \sin 2x\cos x + \cos 2x\sin x$$

We still have $2x$ involved, so need to use double-angle formulae.
We have to decide which to use for $\cos 2x$; looking at the RHS, we want everything in terms of sin, so use $\cos 2x \equiv 1 - 2\sin^2 x$

So $\text{LHS} \equiv \sin 2x\cos x + \cos 2x\sin x \equiv (2\sin x\cos x)\cos x + (1 - 2\sin^2 x)\sin x$
$\equiv 2\sin x\cos^2 x + \sin x - 2\sin^3 x$

This looks closer to the answer – everything is in terms of x. Comparing what we've got to what we want, we can see that everything except the $\cos^2 x$ looks OK – so we need to get rid of this, by using $\sin^2 + \cos^2 = 1$

So: $\text{LHS} \equiv 2\sin x\cos^2 x + \sin x - 2\sin^3 x \equiv 2\sin x(1 - \sin^2 x) + \sin x - 2\sin^3 x$
$\equiv 2\sin x - 2\sin^3 x + \sin x - 2\sin^3 x \equiv 3\sin x - 4\sin^3 x \equiv \text{RHS}$

Putting things in the form $R\sin(x \pm \alpha)$ or $R\cos(x \pm \alpha)$

You are often asked to put expressions like $3\sin x + 4\cos x$ in the form $R\sin(x \pm \alpha)$ (or $R\cos(x \pm \alpha)$ – it works exactly the same way).

Example 4

Express $3\sin x + 4\cos x$ in the form $R\sin(x + \alpha)$, where R is a positive constant and α is an acute angle. Give your answers to one decimal place, where appropriate.

Solution

Step 1: Put the two things equal — $3\sin x + 4\cos x = R\sin(x + \alpha)$

Step 2: Expand — $3\sin x + 4\cos x = R\sin x\cos\alpha + R\cos x\sin\alpha$

Step 3: Equate coefficients — $3 = R\cos\alpha \qquad 4 = R\sin\alpha$

Step 4: Divide the equation with $\sin\alpha$ by the equation with $\cos\alpha$. — $\frac{4}{3} = \frac{R\sin\alpha}{R\cos\alpha} = \tan\alpha \Rightarrow \alpha = 53.1°$

Step 5: Square each of the equations from Step 3, add them up, then use $\sin^2 + \cos^2 = 1$ — $9 = R^2\cos^2\alpha \qquad 16 = R^2\sin^2\alpha$
$25 = R^2(\cos^2\alpha + \sin^2\alpha)$
$25 = R^2 \Rightarrow R = 5$

Be careful to make sure you have the R in both terms

Deciding which form to use

In some applications, you are not told to use $R\sin(x + \alpha)$, say, rather than $R\cos(x + \alpha)$. You then have to decide which to use for yourself. These are the rules:

coefficient of $\sin x$	coefficient of $\cos x$	Use
positive	positive	$R\sin(x + \alpha)$ or $R\cos(x - \alpha)$
positive	negative	$R\sin(x - \alpha)$
negative	positive	$R\cos(x + \alpha)$
negative	negative	$-R\sin(x + \alpha)$ or $-R\cos(x - \alpha)$

Applications

The following examples demonstrate some common applications of this – learn them!

Example 5

Solve the equation $\sqrt{3}\cos x - \sin x = 1 \qquad -90° \leq x \leq 90°$

Solution

Let $\sqrt{3}\cos x - \sin x = R\cos(x + \alpha)$
$\sqrt{3} = R\cos\alpha \qquad 1 = R\sin\alpha$
$\Rightarrow \tan\alpha = \frac{1}{\sqrt{3}} \Rightarrow \alpha = 30° \qquad R = 2$

So $2\cos(x + 30°) = 1$
$\cos(x + 30°) = \frac{1}{2}$
$x = -90°, 30°$

Example 6

Find the range of the function $f(x) = 3 + 2\sin x - \cos x$

Solution

Let $2\sin x - \cos x = R\sin(x - \alpha)$ (ignore the 3 for now!)
$2 - R\cos\alpha \qquad 1 = R\sin\alpha$
$R = \sqrt{5} \qquad \alpha = \tan^{-1}(\frac{1}{2})$

So $f(x) = 3 + \sqrt{5}\sin(x - a)$
Max value of $\sin(x - \alpha)$ is 1, and min is -1
So $3 - \sqrt{5} \leq f(x) \leq 3 + \sqrt{5}$

Trigonometry

Use your knowledge

1 a) Prove $\cos 4x \equiv 8\cos^4 x - 8\cos^2 x + 1$

$\cos 4x \equiv \cos(2x + 2x)$ – use addition formulae
Which formula for $\cos 2x$ is the most useful to use?

b) Hence solve the equation $\cos 4x = \cos^2 x \quad -360° \leq x \leq 360°$

Substitute in for $\cos 4x$ using part a)
Let $y = \cos 2x$. You now have a quadratic
Don't forget $\pm$ square root

c) Write down the solutions to the equation
$\cos 2x = \cos^2(\frac{1}{2}x) \quad -720° \leq x \leq 720°$

Let $x = 2A$
Look back at part b)!

2 Solve the equation $2\cos x\cos\frac{\pi}{3} = 2\sin x\sin\frac{\pi}{3} + 1 \quad -\pi \leq x \leq \pi$

Try taking the $\sin x\sin\pi/3$ over the other side
Try dividing by 2
This should look like a standard formula. Check in your formula book!

3 a) Given that $t = \tan(\frac{1}{2}x)$, show that:

i) $\tan x = \dfrac{2t}{1 - t^2}$ ii) $\sec x = \dfrac{1 + t^2}{1 - t^2}$

In the formula for $\tan 2A$, put $A = \frac{1}{2}x$
Use the relationship between $\sec^2 x$ and $\tan^2 x$

b) Hence solve the equation $\sec x = 2\tan x \tan\frac{1}{2}x \quad -180° < x < 180°$

Put everything in terms of t, then rearrange

4 Prove the identity $\sec x + \tan x \equiv \dfrac{\cos x}{1 - \sin x}$

Put LHS in terms of $\sin x$ and $\cos x$, and put it over a common denominator
Multiply top and bottom by $1 - \sin x$
Use $\sin^2 + \cos^2 = 1$

5 Find the range of the function $f(x) = \dfrac{2}{6 + 4\sin x + 3\cos x}$

Use $R\sin(x + \alpha)$
Find the max and min values of the bottom

Differentiation

Test your knowledge

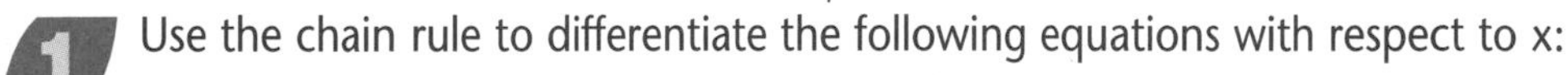

1 Use the chain rule to differentiate the following equations with respect to x:

a) $y = e^{x^2 - 3x + 5}$ b) $y = \ln|\sin x|$ c) $y = 2\sin^{-4}x$

d) $y = \dfrac{1}{\sin x}$ e) $y = (4x^2 + 5x + 2)^{11}$ f) $y = \cos^8 2x$

g) $y = \dfrac{1}{(4x + 3)^2}$ h) $y = \tan(2x - 5)$ i) $y = \ln(x^2 + 5x + 3)$

2 Differentiate the following functions with respect to x:

a) $f(x) = x^2\sin x$ b) $f(x) = e^{2x}\sin x$ c) $f(x) = x^4(x^2 + 7)^5$

3 Differentiate the following equations with respect to x:

a) $y = \dfrac{e^x}{\cos x}$ b) $y = \dfrac{\sin 2x}{x + 5}$ c) $y = \dfrac{\sqrt{x}}{2x^2 + 5}$

4 Calculate the gradient of the curve $y = \dfrac{3x - 1}{(x + 3)(x + 1)}$ at the point $\left(7, \dfrac{1}{4}\right)$.

5 For the curve with equation $y = x\ln x$, find:

a) the exact coordinates of the stationary point

b) the corresponding value of $\dfrac{d^2y}{dx^2}$, and hence state the nature of the stationary point.

6 Find the equations of the tangent and the normal to the curve $y = \dfrac{x}{3x - 2}$, at the point Q, where $x = 2$

7 Find expressions for $\dfrac{dy}{dx}$ in terms of x and y for the following curves that are defined implicitly:

a) $x^2 - \dfrac{y^2}{3} = 1$

b) $16x^2 + 16y^2 - 16x - 24y = 83$

Differentiation

Test your knowledge

8 The length of each side of a large ice cube is decreasing at a constant rate of $0.008\,\text{cms}^{-1}$. Find the rate at which:

a) the volume of the cube
b) the outer surface area of the cube

decrease at the instant when the volume of the cube is $125\,\text{cm}^3$.

9 Find the exact gradients, at the point P, where $x = 7$, for the curves below, which are expressed parametrically:

a) $x = 3t + 1 \qquad y = e^t$
b) $x = 7\sin\theta \qquad y = \cos^2\theta$

Answers

1 a) $(2x - 3)e^{x^2 - 3x + 5}$ **b)** $\frac{\cos x}{\sin x} = \cot x$ **c)** $-8\sin^{-5}x\cos x$ **d)** $-\sin^{-2}x\cos x$
e) $11(8x + 5)(4x^2 + 5x + 2)^{10}$ **f)** $-16\cos^7 2x\sin 2x$ **g)** $\frac{-8}{(4x + 3)^3}$
h) $2\sec^2(2x - 5)$ **i)** $\frac{2x + 5}{x^2 + 5x + 3}$

2 a) $2x\sin x + x^2\cos x$ **b)** $2e^{2x}\sin x + e^{2x}\cos x$ **c)** $4x^3(x^2 + 7)^5 + 10x^5(x^2 + 7)^4 = 14x^3(x^2 + 7)^4(x^2 + 2)$

3 a) $\frac{e^x(\cos x + \sin x)}{\cos^2 x}$ **b)** $\frac{2(x + 5)\cos 2x - \sin 2x}{(x + 5)^2}$ **c)** $\frac{(\frac{5}{2})x^{-\frac{1}{2}} - 3x^{\frac{3}{2}}}{(2x^2 + 5)^2}$

4 $\frac{-3}{160}$

5 a) $(e^{-1}, -e^{-1})$ **b)** $e^1 = e$ minimum SP

6 (T): $y - \frac{1}{2} = -\frac{1}{8}(x - 2)$ (N): $y - \frac{1}{2} = 8(x - 2)$

7 a) $\frac{3x}{y}$ **b)** $\frac{2(1 - 2x)}{(4y - 3)}$

8 a) $0.6\,\text{cm}^3\text{s}^{-1}$ **b)** $0.48\,\text{cm}^2\text{s}^{-1}$

9 a) $\frac{e^2}{3}$ **b)** $-\frac{2}{7}$

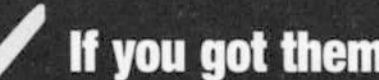
If you got them all right, skip to page 33

Improve your knowledge

1 Basic Differentiation

Below is a table showing basic and general derivatives that are obtained using the chain rule. If the chain rule is a nightmare, learn the general derivatives.

y	$\frac{dy}{dx}$	y	$\frac{dy}{dx}$	y	$\frac{dy}{dx}$
x^n	nx^{n-1}	$(ax + b)^n$	$an(ax + b)^{n-1}$	$(2x + 5)^4$	$8(2x + 5)^3$
e^x	e^x	$e^{ax + b}$	$ae^{ax + b}$	$e^{1 - 2x}$	$-2e^{1 - 2x}$
$\ln\|x\|$	$\frac{1}{x}$	$\ln\|ax + b\|$	$\frac{a}{ax + b}$	$\ln\|5x - 4\|$	$\frac{5}{5x - 4}$
$\sin x$	$\cos x$	$\sin(ax + b)$	$a\cos(ax + b)$	$\sin(7x + 3)$	$7\cos(7x + 3)$
$\cos x$	$-\sin x$	$\cos(ax + b)$	$-a\sin(ax + b)$	$\cos(4 + 3x)$	$-3\sin(4 + 3x)$
$\tan x$	$\sec^2 x$	$\tan(ax + b)$	$a\sec^2(ax + b)$	$\tan(1 - 2x)$	$-2\sec^2(1 - 2x)$

Key points from AS in a Week
Base Differentiation pages 17–18
Tangents and Normals page 18
Stationary Points pages 19–23

In your formula book the chain rule is stated as follows:

$$\boxed{\frac{dy}{dx} = \frac{dy}{du} \times \frac{du}{dx}} \text{ where } y = f(u) \text{ and } u = g(x)$$

This looks scary, but it becomes easier when you look at the next example.

Example 1

Use the chain rule to find $\frac{dy}{dx}$ for the following:

a) $y = \sqrt{(2 - 3x^2)}$ b) $y = \sin^4 x$

Solution

a) Using indices, $y = \sqrt{(2 - 3x^2)} = (2 - 3x^2)^{\frac{1}{2}}$ **(1)**

1) Let u be what's in the brackets: So $u = 2 - 3x^2$

2) Substitute u into **(1)**: So $y = u^{\frac{1}{2}}$, where $u = 2 - 3x^2$

3) Differentiate each: Hence $\frac{dy}{du} = \frac{1}{2}u^{-\frac{1}{2}}$ and $\frac{du}{dx} = -6x$

4) Use formula $\frac{dy}{dx} = \frac{dy}{du} \times \frac{du}{dx}$: So $\frac{dy}{dx} = \left(\frac{1}{2}u^{-\frac{1}{2}}\right) \times (-6x) = -3xu^{-\frac{1}{2}}$

5) Substituting back: $\frac{dy}{dx} = -3x(2 - 3x^2)^{-\frac{1}{2}} = \frac{-3x}{(2 - 3x^2)^{\frac{1}{2}}}$

b) $y = \sin^4 x = (\sin x)^4$ That's u!

1 and 2) $y = u^4$ where $u = \sin x$

3) $\frac{dy}{du} = 4u^3$ and $\frac{du}{dx} = \cos x$

4 and 5) $\frac{dy}{dx} = 4u^3 \cos x = 4\sin^3 x \cos x$

If you want to save time and effort in an exam, it is good practice to learn the generalised form of the chain rule, as applied in the examples in the table below.

y	$\frac{dy}{dx}$	y	$\frac{dy}{dx}$
$(f(x))^n$	$nf'(x)(f(x))^{n-1}$	$(4x^2 - 7)^5$	$5 \times 8x(4x^2 - 7)^4$ $= 40x(4x^2 - 7)^4$
$e^{f(x)}$	$f'(x)e^{f(x)}$	$e^{-7x^3 + 5x}$	$(-21x^2 + 5)e^{-7x^3 + 5x}$
$\ln f(x)$	$\frac{f'(x)}{f(x)}$	$\ln \cos x$	$\frac{-\sin x}{\cos x} = -\tan x$
$\sin(f(x))$	$f'(x)\cos(f(x))$	$\sin(3x^3)$	$9x^2\cos(3x^3)$
$\cos(f(x))$	$-f'(x)\sin(f(x))$	$\cos(e^{\frac{x}{2}})$	$-\frac{1}{2}e^{\frac{x}{2}}\sin(e^{\frac{x}{2}})$
$\tan(f(x))$	$f'(x)\sec^2(f(x))$	$\tan(1 + 4x^2)$	$8x\sec^2(1 + 4x^2)$
$\sin^n x$	$n\sin^{n-1}x\cos x$	$\sin^4 x$	$4\sin^3 x\cos x$
$\cos^n x$	$-n\cos^{n-1}x\sin x$	$\cos^5 x$	$-5\cos^4 x\sin x$

2 Product Rule ($y = u.v$)

When two separate functions with x in them are multiplying each other, we use the product rule. We call one of the products u, the other v.

When: $y = u.v$ then: $\frac{dy}{dx} = v.u' + u.v'$

This formula is found in your formula book!

There are four ingredients to help us to find $\frac{dy}{dx}$. They are u, v, $u' = \frac{du}{dx}$ and $v' = \frac{dv}{dx}$

Example 2

Find $\frac{dy}{dx}$ for the function $y = x^2\cos 3x$

Solution

Let: $u = x^2 \qquad v = \cos 3x$

Differentiate each: $u' = 2x \qquad v' = -3\sin 3x$

Use formula: $\frac{dy}{dx} = 2x\cos 3x - 3x^2\sin 3x$

3 Quotient Rule ($y = u/v$)

To use the quotient rule we need one function divided by another. The one at the top is called u, the bottom one v.

When: $y = \frac{u}{v}$ then: $\boxed{\frac{dy}{dx} = \frac{v.u' - u.v'}{v^2}}$

Your formula book is there to help!!!

Again, for the quotient rule, you need to find the same four magic ingredients and insert them into the formula.

Example 3

Find the gradient of the curve $y = \frac{e^{2x}}{3x + 5}$, when $x = 0$.

Solution

Functions dividing ⇒ Quotient Rule.

Let $u = e^{2x} \qquad v = 3x + 5$

Differentiate: $u' = 2e^{2x} \qquad v' = 3$

Formula: $\frac{dy}{dx} = \frac{2(3x + 5)e^{2x} - 3e^{2x}}{(3x + 5)^2} = \frac{6xe^{2x} + 10e^{2x} - 3e^{2x}}{(3x + 5)^2}$

giving: $\frac{dy}{dx} = \frac{(6x + 7)e^{2x}}{(3x + 5)^2}$

Noting that $\frac{dy}{dx}$ also means the gradient of the curve,

when $x = 0$, gradient $= \frac{dy}{dx} = \frac{(6 \times 0 + 7)e^0}{(3 \times 0 + 5)^2} = \frac{7 \times 1}{5^2} = \frac{7}{25}$

Partial Fractions

Some expressions must be written as a partial fraction before they can be differentiated. Remember to look out for these!

Example 4

If $f(x) = \dfrac{x^2 - 9x + 17}{(x-2)^2(x+1)}$ then find $f'(x)$.

Solution

1) Express $f(x)$ in partial fractions:

$f(x) = \dfrac{1}{(x-2)^2} - \dfrac{2}{(x-2)} + \dfrac{3}{(x+1)}$ when expressed in partial fraction form.

2) Express $f(x)$ in indices form: $f(x) = (x-2)^{-2} - 2(x-2)^{-1} + 3(x+1)^{-1}$

3) Differentiate using the chain rule: $f'(x) = -2(x-2)^{-3} + 2(x-2)^{-2} - 3(x+1)^{-2}$

$f'(x)$ may also be written as: $f'(x) = -\dfrac{2}{(x-2)^3} + \dfrac{2}{(x-2)^2} - \dfrac{3}{(x+1)^2}$.

5 Stationary Points

You still need to remember the work done on stationary points in AS Maths and apply the theory to more difficult expressions found in A2 Maths.

Example 5

For the curve $y = \dfrac{\ln x}{x}$, find: a) the exact coordinates of the stationary point;

b) the corresponding value of $\dfrac{d^2y}{dx^2}$. Finally, c) state the nature of the SP.

Solution

a) 1) Differentiate: The two functions are dividing, so use the quotient rule: $u = \ln x$ $\quad v = x$

$u' = \dfrac{1}{x}$ $\quad v' = 1 \Rightarrow \dfrac{dy}{dx} = \dfrac{1 - \ln x}{x^2}$

2) Set $\dfrac{dy}{dx} = 0$ to solve for x: $\dfrac{1 - \ln x}{x^2} = 0$ $\quad (\times\, x^2) \Rightarrow 1 - \ln x = 0$

$\Rightarrow 1 = \ln x \Rightarrow x = e^1 = e$

3) Find y-coordinate $\quad y = \dfrac{\ln e}{e} = \dfrac{1}{e} = e^{-1}$ $\quad (\because \ln e = 1)$

Hence the coordinates of the SP are $\left(e, \dfrac{1}{e}\right)$.

b) Quotient rule again with: $u = 1 - \ln x \quad v = x^2$

$u' = -\frac{1}{x} \qquad v' = 2x$

giving $\frac{d^2y}{dx^2} = \frac{-x - 2x(1 - \ln x)}{x^4} = \frac{-3x + 2x\ln x}{x^4} = \frac{-3 + 2\ln x}{x^3}$

When $x = e$, $\frac{d^2y}{dx^2} = \frac{-3 + 2\ln e}{e^3} = \frac{-3 + 2}{e^3} = -\frac{1}{e^3}$.

c) Since $\frac{d^2y}{dx^2} = -\frac{1}{e^3} < 0$ we conclude there is a maximum SP at $\left(e, \frac{1}{e}\right)$

6 Tangents and Normals

Again, you need to work out the equations of tangents and normals in A2 Maths, as in AS Maths.

Example 6

Work out (in terms of e) the equations of the tangent and the normal to the curve $y = x^2e^{2x}$, at the point Q, where $x = \frac{1}{2}$.

Solution

Point Q $\Rightarrow y = (\frac{1}{2})^2e^{2(0.5)} = \frac{1}{4}e \Rightarrow Q(\frac{1}{2}, \frac{1}{4}e)$.

Gradient $\Rightarrow \frac{dy}{dx}$ $\qquad u = x^2 \qquad v = e^{2x}$

Differentiate using Product Rule. $\qquad u' = 2x \qquad v' = 2e^{2x}$

$\Rightarrow \frac{dy}{dx} = 2xe^{2x} + 2x^2e^{2x}$

NB: $\frac{dy}{dx}$ is the same as the gradient of the tangent.

So at Q, gradient of (T) $= \frac{dy}{dx} = 2\left(\frac{1}{2}\right)e^1 + 2\left(\frac{1}{2}\right)^2e^1 = \frac{3}{2}e$

A gradient and a point makes a line

Gradient and point $\Rightarrow$ Equation of (T) is: $y - \frac{1}{4}e = \frac{3}{2}e\left(x - \frac{1}{2}\right)$

Gradient of normal $= \frac{-1}{\text{Gradient of tangent}} \Rightarrow$ Gradient (N) $= \frac{-1}{\frac{3e}{2}} = \frac{-2}{3e}$

Gradient and point $\Rightarrow$ Equation of (N) is: $y - \frac{1}{4}e = \frac{-2}{3e}\left(x - \frac{1}{2}\right)$

Implicit Differentiation

So far we have only differentiated explicit equations like $y = x^2 \cos 3x$, $y = x^2 - 2x$, etc., which are written in the form: $y = f(x)$ (i.e. y = something in terms of x). We use implicit differentiation, to differentiate equations like $9x^2 + 4y^2 = 36$ or $e^{2x} \ln y - x = 0$, which are expressed implicitly in the form: $f(x, y)$ = constant.

Example 7

Find $\frac{dy}{dx}$ in terms of x and y for each of the following equations:

a) $9x^2 + 4y^2 = 36$ b) $x^2 + y^2 - 2x + 4y = 5$

Solution

a) Differentiate each term wrt x: i.e. $\frac{d}{dx}(9x^2) + \frac{d}{dx}(4y^2) = \frac{d}{dx}(36)$

Note that: $\frac{d}{dx}(9x^2) = 18x$ and $\frac{d}{dx}(36) = 0$

However: $\frac{d}{dx}(4y^2) = \frac{d}{dy}(4y^2)\frac{dy}{dx} = 8y\frac{dy}{dx}$

$\frac{d}{dx}$ means differentiate with respect to x

Rule: If you differentiate a y-term with respect to x, then you differentiate it with respect to y, and times the answer by $\frac{dy}{dx}$

Continuing with solution: $18x + 8y\frac{dy}{dx} = 0$

Rearranging: $8y\frac{dy}{dx} = -18x \Rightarrow \frac{dy}{dx} = \frac{-18x}{8y}$

Answer: $\frac{dy}{dx} = \frac{-9x}{4y}$

b) Differentiating wrt x: $\frac{d}{dx}(x^2) + \frac{d}{dx}(y^2) - \frac{d}{dx}(2x) + \frac{d}{dx}(4y) = \frac{d}{dx}(5)$

Hence $2x + 2y\frac{dy}{dx} - 2 + 4\frac{dy}{dx} = 0$

Rearranging: $(2y + 4)\frac{dy}{dx} = 2 - 2x$

$$\frac{dy}{dx} = \frac{2 - 2x}{2y + 4} = \frac{2(1 - x)}{2(y + 2)}$$

Answer: $\frac{dy}{dx} = \frac{1 - x}{y + 2}$

The problem encountered with implicit differentiation is when you are asked to differentiate an expression that is a product of an x and a y term. To do this, we use the product rule.

Eg: $\frac{d}{dx}(x^3\ln y) = v.u' + u.v' = \ln y(3x^2) + x^3\frac{1}{y}\frac{dy}{dx}$ (where $u = x^3$, $v = \ln y$)

$$= 3x^2\ln y + \frac{x^3}{y}\frac{dy}{dx}$$

Example 8

Find the equation of the normal to the curve $6x^2 + y^2 - 2xy - 6 = 0$, at the point Q with coordinates (1, 2).

Solution

Need to find $\frac{dy}{dx}$: $\frac{d}{dx}(6x^2) + \frac{d}{dx}(y^2) - \frac{d}{dx}(2xy) - \frac{d}{dx}(6) = \frac{d}{dx}(0)$

Product rule: $\frac{d}{dx}(2xy) = y2 + 2x1\frac{dy}{dx} = 2y + 2x\frac{dy}{dx}$

Differentiating: $12x + 2y\frac{dy}{dx} - \left(2y + 2x\frac{dy}{dx}\right) - 0 = 0$

Rearranging: $(2y - 2x)\frac{dy}{dx} = -12x + 2y$

Hence: $\frac{dy}{dx} = \frac{-6x + y}{y - x}$

Gradient of Tangent at Q(1, 2): $\frac{dy}{dx} = \frac{-6(1) + (2)}{2 - 1} = -4$

Hence, gradient of (N) at Q is: $\frac{1}{4}$ and point is (1, 2)

Equation of Normal: $y - 2 = \frac{1}{4}(x - 1)$

8 Connected Rates of Change

In this topic we may know the rate of change of one quantity, such as the area of a circle (i.e. $\frac{dA}{dt}$), and we may need to find the rate of change of a related quantity such as its radius (i.e. $\frac{dr}{dt}$).

If a quantity is increasing, then the rate of change is positive; but if it's decreasing, then the rate of change is negative.

Example 9

The area of a circular stain is increasing at a rate of $2\,\text{cm}^2\,\text{s}^{-1}$. Find the rate at which the radius of the circle is increasing when the area of the circle is $4\pi\,\text{cm}^2$.

Solution

Let r = radius, A = area of circle. We know $\frac{dA}{dt} = 2\,\text{cm}^2\text{s}^{-1}$

rate at which radius is increasing $\Rightarrow$ we need to find $\frac{dr}{dt}$

We link $\frac{dr}{dt}$ to $\frac{dA}{dt}$, by the chain rule: i.e. $\frac{dr}{dt} = \frac{dr}{dA} \times \frac{dA}{dt}$ **(2)**

$\frac{dr}{dA}$ is the missing link $\Rightarrow$ we need to find a relationship between A and r

Since the stain is a circle, then: $A = \pi r^2 \Rightarrow \frac{dA}{dr} = 2\pi r \Rightarrow \frac{dr}{dA} = \frac{1}{2\pi r}$

Hence: $\frac{dr}{dt} = \frac{1}{2\pi r} \times 2 = \frac{1}{\pi r}$ by **(2)** $\Rightarrow$ we need to find r

Find r: $A = \pi r^2 = 4\pi \Rightarrow r^2 = 4 \Rightarrow r = 2\,\text{cm}$

$\therefore \frac{dr}{dt} = \frac{1}{\pi(2)} = 0.15915... = 0.159\,\text{cm s}^{-1}$

Note: If the units are given in a question, you should give the appropriate units in your answer. The units are also a great help; i.e. $\text{cm}^2\text{s}^{-1} \Rightarrow$ area per time $\Rightarrow \frac{dA}{dt}$

9 Parametric Differentiation

Parametric equations are where the variables x and y are expressed in terms of another variable, say t. The first derivative for parametric functions is given by the formula:

$$\boxed{\frac{dy}{dx} = \frac{dy}{dt} \bigg/ \frac{dx}{dt}}$$

Example 10

Find the gradient of the curve: $x = 4\sin t$ $\quad y = 3\cos t$, when $t = \frac{\pi}{4}$

Solution

Differentiate each: $\frac{dx}{dt} = 4\cos t$ $\quad \frac{dy}{dt} = -3\sin t$

Use formula $\frac{dy}{dt} = \frac{-3\sin t}{4\cos t} = -\frac{3}{4}\tan t$ $\quad \because \frac{\sin t}{\cos t} = \tan t$

When $t = \frac{\pi}{4}, \frac{dy}{dx} = -\frac{3}{4}\tan\left(\frac{\pi}{4}\right) = -\frac{3}{4}$

Example 11

A curve is defined parametrically by the equations:

$x = t^3 - 5t, \qquad y = t^2$ **(3)**

a) Calculate the equation of: i) the tangent T and

ii) the normal N to the curve at the point P, where $t = 2$

b) Deduce that the tangent intersects the curve when $4t^3 - 7t^2 - 20t + 36 = 0$ **(4)**

c) Calculate the coordinates of Q, where the tangent cuts the curve again.

Solution

a) i) Point: $t = 2 \Rightarrow x = 8 - 10 = -2, \qquad y = 2^2 = 4.$ So P(−2, 4)

Gradient: $\frac{dx}{dt} = 3t^2 - 5$ and $\frac{dy}{dt} = 2t \Rightarrow \frac{dy}{dx} = \frac{2t}{3t^2 - 5}$, parametrically

Gradient at (T): $\frac{dy}{dx} = \frac{2(2)}{3(4) - 5} = \frac{4}{7}$

Hence equation of (T): $y - 4 = \frac{4}{7}(x + 2)$ giving: $y = \frac{4}{7}x + \frac{36}{7}$

ii) Gradient (N) $= \frac{-1}{\text{Gradient (T)}} = \frac{-1}{4/7} = -\frac{7}{4}$

Hence equation of (N): $y - 4 = -\frac{7}{4}(x + 2)$ giving: $y = -\frac{7}{4}x + \frac{1}{2}$

b) Tangent **(T)** intersects curve **(3)** $\Rightarrow$ Solve these equations simultaneously

Substituting **(3)** into **(T)** gives: $t^2 = \frac{4}{7}(t^3 - 5t) + \frac{36}{7}$

Multiplying both sides by 7: $7t^2 = 4t^3 - 20t + 36$

Rearranging gives: $4t^3 - 7t^2 - 20t + 36 = 0$, as required **(4)**

c) We need to find where the tangent cuts the curve again.

To do this we solve the equation **(4)**, to find its roots (i.e. solutions for t).

Solutions to equation **(4)** tell us where the curve meets the tangent.

We know the tangent *touches* the curve when $t = 2$. This is a solution to **(4)**.

$t = 2$ is a solution $\Rightarrow (t - 2)$ is a factor, by the factor theorem.

Trick: Because the tangent touches the curve, the root is a repeated one.

Hence $(t - 2)^2$ is a repeated factor.

$\therefore 4t^3 - 7t^2 - 20t + 36 = 0 = (t - 2)^2(\text{something}) = (t^2 - 4t + 4)(4t + 9)$

So **(4)** $= (t - 2)^2(4t + 9)$. Then $t = 2$, which we already know, and $4t + 9 = 0$.

Hence $t = -\frac{9}{4}$ with $x = \left(\frac{-9}{4}\right)^3 - 5\left(\frac{-9}{4}\right) = -\frac{9}{64}$ and $y = \left(\frac{-9}{4}\right)^2 = \frac{81}{16}$

$\therefore$ the coordinates of intersection with the curve is $Q\left(-\frac{9}{64}, \frac{81}{16}\right)$

Use your knowledge

1 A curve is given by the equation, $y = (x + 2)e^{-2x}$

a) Find $\frac{dy}{dx}$

Use product rule to differentiate

Factorise out e^{-2x}, to make later working easier!

b) Find the exact coordinates of the stationary point C.

Set (a) to zero. Calculate x, then y. NB: $e^{-2x} = 0$, has no solutions

c) Work out $\frac{d^2y}{dx^2}$ at the point C. Hence determine the nature of the stationary point.

Product Rule again! Apply 2nd derivative test

2 A curve is given parametrically by the equations:

$x = 2t \qquad y = \frac{6}{t}, \qquad t \in \mathbb{R}, \qquad t \neq 0$

a) Find $\frac{dy}{dx}$, in terms of t

Use parametric differentiation

The normal to curve is drawn at the point P, where $t = 3$

b) Show that the equation of the normal is $y = 3x - 16$

Find gradient of tangent. Use $m_1.m_2 = -1$ to find normal

The normal meets the curve again at the point Q.

c) Calculate the coordinates of the point Q.

Make normal = curve. Solve quadratic for t. Find coordinates

3 A curve is defined implicitly by the equation:

$y^3 - 6x + 2xy - 10 = 0$

Use implicit differentiation

a) Find $\frac{dy}{dx}$, in terms of x and y

Watch out for the product rule! Then rearrange!

b) Hence, calculate the equation of the tangent at the point Q, with y-coordinate 2.

Find the x-value and gradient. Gradient & Point makes a line

4 Find the set of values for x, for which the curve with equation $y = \frac{e^{2x}}{(x + 3)}$, $x \neq -3$ is increasing.

Differentiate using quotient rule. Increasing function $\Rightarrow \frac{dy}{dx} > 0$

Integration

Test your knowledge

1 Integrate the following with respect to x:

a) $\dfrac{2}{(5x-3)^3}$ b) $(3-4x)^5$ c) $\dfrac{1}{3x+5}$ d) $\dfrac{2}{e^{1-2x}}$

e) $(e^{-x}+2)^2$ f) e^{4-x} g) $\cos(1-2x)$ h) $-3\sec^2(2x)$

2 Find $\displaystyle\int \frac{3x+5}{(x+1)(x+3)}\,dx$

3 a) Differentiate $\sqrt{7+2x^2}$, with respect to x

b) Using the result from (a), evaluate $\displaystyle\int_1^3 \frac{x}{\sqrt{7+2x^2}}\,dx$

4 Find, using appropriate trigonometric identities:

a) $\int \cos^2 x dx$, and evaluate $\int_0^{\frac{\pi}{3}} \cos^2 x dx$

b) $\int \sin 2x \sin 4x\, dx$ (Hint: Write the other way around!)

c) $\int \tan^2 x\, dx$

5 Find, using appropriate substitutions:

a) $\int x(3x^2-5)^7 dx$, $(u = 3x^2 - 5)$

b) $\int_0^{\frac{\pi}{6}} \sin^2 x \cos x dx$, $(u = \sin x)$

c) $\sin^3 x dx$

d) $\int 4x\sqrt{(3-4x)}dx$

6 Find:

a) $\displaystyle\int \frac{3x^2+5}{x^3+5x-8}\,dx$ b) $\displaystyle\int \frac{x+2}{x^2+4x-10}\,dx$ c) $\displaystyle\int \frac{3x+6}{x^2+4x+5}\,dx$

7 Find:

a) $\int x \sin x dx$ b) $\int x^2 e^{3x} dx$

Test your knowledge

8 a) Find the particular solution of the differential equation:

$x\dfrac{dy}{dx} = y^2$; given that when $x = 1$, $y = 1$

b) Find the general solution of the differential equation:

$\dfrac{dy}{dx} = \dfrac{1}{e^{-2x+y}}$ leaving your answer in the form of exponentials.

c) Find the general solution, in the form $y = f(x)$, of the differential equation:

$(1 + x^3)\dfrac{dy}{dx} = x^2y$

9 Calculate the solid volume of revolution formed when the region bounded by the curve $y = 2\ln x$, the lines $y = 1$ and $y = 3$ is rotated 2π radians about the y-axis. Give your answer to 3 significant figures.

Answers

1 a) $-\frac{1}{5}(5x-3)^{-2} + c$ **b)** $-\frac{1}{24}(3-4x)^6 + c$ **c)** $\frac{1}{3}\ln|3x+5| + c$
d) $e^{-1+2x} + c$ **e)** $-\frac{1}{2}e^{-2x} - 4e^{-x} + 4x + c$ **f)** $-e^{4-x} + c$
g) $-\frac{1}{2}\sin(1-2x) + c$ **h)** $-\frac{3}{2}\tan 2x + c$

2 $\ln(x+1) + 2\ln(x+3) + c = \ln(x+1)(x+3)^2 + c$

3 a) $\frac{2x}{\sqrt{(7-2x^2)}}$ **b)** 1

4 a) $\frac{1}{4}\sin 2x + \frac{1}{2}x + c$; $\frac{\sqrt{3}}{8} + \frac{\pi}{6}$ **b)** $\frac{1}{4}\sin 2x - \frac{1}{12}\sin 6x + c$ **c)** $\tan x - x + c$

5 a) $\frac{1}{48}(3x^2-5)^8 + c$ **b)** $\frac{1}{24}$ **c)** $\frac{1}{3}\cos^3 x - \cos x + c$
d) $-\frac{1}{2}(3-4x)^{\frac{3}{2}} + \frac{1}{10}(3-4x)^{\frac{5}{2}} + c$

6 a) $\ln|x^3 + 5x - 8| + c$ **b)** $\frac{1}{2}\ln|x^2 + 4x - 10| + c$ **c)** $\frac{3}{2}\ln|x^2 + 4x + 5| + c$

7 a) $-x\cos x + \sin x + c$ **b)** $\frac{1}{3}x^2e^{3x} - \frac{2}{9}xe^{3x} + \frac{2}{27}e^{3x} + c$

8 a) $-\frac{1}{y} = \ln x - 1$ or $y = \frac{1}{1-\ln x}$ **b)** $e^y = \frac{1}{2}e^{2x} + c$ **c)** $y = A\sqrt[3]{(1+x^3)}$

9 $54.6(\text{units})^3$

If you got them all right, skip to page 50

Integration

Improve your knowledge

1 Standard Linear Integrals

A lot of valuable time can be saved when we have to integrate basic functions that are expressed in a more general linear form. For example, instead of integrating e^x or sin x, we have to integrate e^{2x+5} or sin $(ax + b)$. We call these standard integrals and these must be learnt.

Key points from AS in a Week

Review the chapter on Integration pages 25–30

General case | **Example**

$f(x)$	$\int f(x)dx$	$f(x)$	$\int f(x)dx$
$(\boldsymbol{ax+b})^n$	$\frac{(\boldsymbol{ax+b})^{n+1}}{a(n+1)}$	$(\boldsymbol{7x-3})^{11}$	$\frac{1}{84}(\boldsymbol{7x-3})^{12}+c$
$e^{\boldsymbol{ax+b}}$	$\frac{1}{a}e^{\boldsymbol{ax+b}}+c$	$e^{\boldsymbol{1-2x}}$	$-\frac{1}{2}e^{\boldsymbol{1-2x}}+c$
$\frac{1}{\boldsymbol{ax+b}}$	$\frac{1}{a}\ln\lvert\boldsymbol{ax+b}\rvert$	$\frac{1}{\boldsymbol{3-4x}}$	$-\frac{1}{4}\ln\lvert\boldsymbol{3-4x}\rvert+c$
$\sin(\boldsymbol{ax+b})$	$-\frac{1}{a}\cos(\boldsymbol{ax+b})+c$	$\sin\boldsymbol{2x}$	$-\frac{1}{2}\cos\boldsymbol{2x}+c$
$\cos(\boldsymbol{ax+b})$	$\frac{1}{a}\sin(\boldsymbol{ax+b})+c$	$\cos(\boldsymbol{1-5x})$	$-\frac{1}{5}\sin(\boldsymbol{1-5x})+c$
$\sec^2(\boldsymbol{ax+b})$	$\frac{1}{a}\tan(\boldsymbol{ax+b})+c$	$\sec^2(\boldsymbol{-4x+3})$	$-\frac{1}{4}\tan(\boldsymbol{-4x+3})+c$

It is important to note that the standard integrals above only work if the term as shown in bold is linear. 'Linear' means that the highest power of x is 1, so terms like $2x - 3$, $1 - 5x$ or $7x$ are allowed; but not $2x^2 + 3$ nor $x^3 - 7x - 2$.

2 Partial Fractions

Some exam questions will ask you to integrate expressions that need to be expressed in partial fractions before they can even be integrated.

Example 1

Show that $\displaystyle\int_{2.5}^{5} \frac{7x^2 - 14x + 6}{(2x-1)(x-1)^2}\,dx = \ln 24 - \frac{5}{12}$

Solution

Express as partial fraction: $\dfrac{7x^2 - 14x + 6}{(2x-1)(x-1)^2} = \dfrac{A}{2x-1} + \dfrac{B}{x-1} + \dfrac{C}{(x-1)^2}$

giving $7x^2 - 14x + 6 = A(x - 1)^2 + B(2x - 1)(x - 1) + C(2x - 1)$

Let $x = 1 \Rightarrow 7 - 14 + 6 = 0 + 0 + C(1) \Rightarrow C = -1$

Let $x = 0.5 \Rightarrow \frac{7}{4} - 7 + 6 = \frac{1}{4}A \Rightarrow \frac{3}{4} = \frac{1}{4}A \Rightarrow A = 3$

Try $x = 0 \Rightarrow 0 - 0 + 6 = A(1)^2 + B(-1)(-1) + C(-1) \Rightarrow 6 = A + B - C$

Substituting for A and C gives $6 = 3 + B + 1 \Rightarrow B = 2$

So $\int_{2.5}^{5} \frac{7x^2 - 14x + 6}{(2x - 1)(x - 1)^2}\, dx = \int_{2.5}^{5} \left(\frac{3}{2x - 1} + \frac{2}{x - 1} - \frac{1}{(x - 1)^2}\right) dx$

Let's try to integrate! The first two terms are ln integrals because they are of the form of $\frac{number}{(ax + b)}$, and the final term can be expressed in the form $(ax + b)^n$.

$$\int_{2.5}^{5} \left(\frac{3}{2x - 1} + \frac{2}{x - 1} - (x - 1)^{-2}\right) dx$$

$$= \left[\frac{3}{2}\ln|2x - 1| + 2\ln|x - 1| - \frac{(x - 1)^{-1}}{-1}\right]_{2.5}^{5}$$

$$= \left[\frac{3}{2}\ln|2x - 1| + 2\ln|x - 1| + \frac{1}{(x - 1)}\right]_{2.5}^{5}$$

$$= \left(\frac{3}{2}\ln 9 + 2\ln 4 + \frac{1}{4}\right) - \left(\frac{3}{2}\ln 4 + 2\ln\frac{3}{2} + \frac{1}{\left(\frac{3}{2}\right)}\right)$$

$$= \left(\ln 9^{\frac{3}{2}} + \ln 4^2 + \frac{1}{4}\right) - \left(\ln 4^{\frac{3}{2}} + \ln\left(\frac{3}{2}\right)^2 + \frac{2}{3}\right)$$

$$= \left(\ln 27 + \ln 16 + \frac{1}{4}\right) - \left(\ln 8 + \ln\frac{9}{4} + \frac{2}{3}\right)$$

$$= \left(\ln 27 + \ln 16 + \frac{1}{4}\right) - \left(\ln 8 + \ln 9 - \ln 4 + \frac{2}{3}\right)$$

$$= \ln 27 + \ln 16 - \ln 8 - \ln 9 + \ln 4 + \frac{1}{4} - \frac{2}{3}$$

$$= \ln\left(\frac{27 \times 16 \times 4}{8 \times 9}\right) - \frac{5}{12} = \ln\left(\frac{1728}{72}\right) - \frac{5}{12}$$

$= \ln 24 - \frac{5}{12}$, as required.

Laws of Logarithms:
$b.\ln a = \ln a^b$
$\ln a + \ln b = \ln(ab)$
$\ln a - \ln b = \ln\left(\frac{a}{b}\right)$

To do definite integration with partial fractions, you can see that it is important to be familiar with the rules of logarithms.

3 Integration as the Reverse of Differentiation

Some questions may give you something to differentiate in the first part, and then ask you to integrate an expression that looks similar to the answer of the first part.

Example 2

a) Differentiate $(9 + 2x^3)^{\frac{1}{2}}$ with respect to x

b) Using the result from a), evaluate $\displaystyle\int_0^2 \frac{x^2}{\sqrt{(9 + 2x^3)}}\,dx$

Solution

a) We use the quick version of the chain rule to differentiate this:

$$\frac{dy}{dx} = \frac{1}{2}(6x^2)\,(9 + 2x^3)^{-\frac{1}{2}} = 3x^2(9 + 2x^3)^{-\frac{1}{2}} = \frac{3x^2}{\sqrt{(9 + 2x^3)}}$$

b) The answer to part a) looks like the integral in part b), except for the number 3. So part (b) can be written as:

$$\frac{1}{3}\int_0^2 \frac{3x^2}{\sqrt{(9 + 2x^3)}}\,dx \qquad \textbf{(1)}$$

The $\frac{1}{3}$ is there to make the integral multiply out to what's asked in (b)

Now the integral is the same as the answer to part a).

Using the fact that integration is the reverse of integration, the integral becomes:

$$= \frac{1}{3}\left[(9 + 2x^3)^{\frac{1}{2}}\right]_0^2 = \frac{1}{3}\left((9 + 16)^{\frac{1}{2}} - (9 + 0)^{\frac{1}{2}}\right) = \frac{1}{3}(5 - 3) = \frac{2}{3}$$

4 Using Trigonometric Identities

When doing integration it is important that you are aware of these three trigonometric identities that you have seen in the trigonometry chapter:

$\sin^2 x + \cos^2 x = 1$ $\qquad \cos 2x = 2\cos^2 x - 1$ $\qquad \cos 2x = 1 - 2\sin^2 x$

Example 3

(A classic integral) Find $\int \sin^2 x\, dx$

Solution

This seems easy, but it is not! The only way you can integrate $\sin^2 x$ (or even $\cos^2 x$) is by using the double-angle formula for cos. There's no other way, so LEARN it!

There are three versions of the double-angle formula for cos. Since we are integrating $\sin^2 x$, we choose the version containing only sine squares.

Choose the appropriate identity: $\cos 2x = 1 - 2\sin^2 x$

Rearrange to make $\sin^2 x$ the subject: $\sin^2 x = \frac{1}{2}(1 - \cos 2x)$

Substituting for $\sin^2 x$ gives: $\int \frac{1}{2}(1 - \cos 2x)dx$

Then integrate: $= \frac{1}{2}(x - \frac{1}{2}\sin 2x) + c$

Simplifying gives: $= \frac{1}{2}x - \frac{1}{4}\sin 2x + c$

Example 4

Use appropriate trigonometry identities, to find:

a) $\int \sin 5x \cos 3x \, dx$ b) $\int \tan^2 3x \, dx$

Solution

a) A product $\Rightarrow$ use trigonometric product formula:

Hence: $\sin A + \sin B = 2\sin\left(\frac{A+B}{2}\right)\cos\left(\frac{A-B}{2}\right) = 2\sin 5x \cos 3x$

Compare terms: $\left(\frac{A+B}{2}\right) = 5x$ and $\left(\frac{A-B}{2}\right) = 3x \Rightarrow A + B = 10x$ and $A - B = 6x$

Solve to find A and B: $\Rightarrow A = 8x$ and $B = 2x$

Identity $\Rightarrow 2\sin 5x\cos 3x = \sin 8x + \sin 2x \Rightarrow \sin 5x \cos 3x = \frac{1}{2}(\sin 8x + \sin 2x)$

Integrate: $\int \sin 5x \cos 3x = \int \frac{1}{2}(\sin 8x + \sin 2x)dx = \frac{1}{2}\left(-\frac{\cos 8x}{8} - \frac{\cos 2x}{2}\right) + c$

b) We cannot integrate $\tan^2 A$; but we can integrate $\sec^2 A$

Identity: $\sec^2 A = \tan^2 A + 1 \Rightarrow \sec^2 3x = \tan^2 3x + 1$, where $A = 3x$

Integral: $\int \tan^2 3x \, dx = \int (\sec^2 3x - 1) \, dx = \frac{1}{3}\tan 3x - x + c$

5 Integration by Substitution

This is a way in which a difficult integral can be broken down into an easier form, which can easily be integrated using the standard methods. The question is usually posed in one variable, say x, and by making a substitution, we completely change the integral to be expressed in a different variable, say u. You are usually (but not always) given the substitutions in examination questions.

Example 5

Find $\int x^2(2x^3 + 7)^5\, dx$ using the substitution $u = 2x^3 + 7$ **(2)**

Solution

1) Differentiate the substitution: $\dfrac{du}{dx} = 6x^2$

2) Rearrange the differential: $\dfrac{du}{6} = x^2\, dx$ **(3)**

OBJECTIVE: We hate and want to get rid of the x's but we now love the u's.

Using equations (2) and (3), we can eliminate the x's.

3) Make the substitutions (2) and (3) Then look – no x's! $\int x^2(2x^3 + 7)^5\, dx = \int u^5 \dfrac{du}{6}$

4) Integrate with respect to u: $= \dfrac{1}{6}\left(\dfrac{u^6}{6}\right) + c$

5) Substitute back for x $= \dfrac{1}{36}(2x^3 + 7)^6 + c$

For definite integrals, we need to change the limits when we use integration by substitution. This is seen in the next example.

Example 6

Evaluate $\int_{\frac{\pi}{6}}^{\frac{\pi}{2}} \sin^3 x \cos x dx$, using the substitution $u = \sin x$

Solution

1) Differentiate substitution: $\dfrac{du}{dx} = \cos x$

2) Rearrange the differential: $du = \cos x\, dx$

3) Change Limits, from x to u When $x = \left(\dfrac{\pi}{2}\right)$, $u = \sin\left(\dfrac{\pi}{2}\right) = 1$

using the substitution formula When $x = \left(\dfrac{\pi}{6}\right)$, $u = \sin\left(\dfrac{\pi}{6}\right) = \dfrac{1}{2}$

4) Make the substitutions, remembering to change your limits $= \displaystyle\int_{0.5}^{1} u^3 du$

5) Integrate with respect to u $= \left[\frac{u^4}{4}\right]_{0.5}^{1}$

6) Evaluate $= \left(\frac{1}{4} - \frac{1}{64}\right) = \frac{15}{64}$

When applying definite integrals for trigonometric functions, the limits must always be in radians.

Example 7

(Another classic integral) Find $\int \cos^3 x \, dx$

Solution

The only way you can integrate $\cos^3$ (or even $\sin^3$) is by using the trigonometric identity $\cos^2 x + \sin^2 x = 1$, followed by integration by substitution.

1) Split up integral $\int \cos^3 x \, dx = \int \cos^2 x \cos x \, dx$
2) Use the identity $\cos^2 x = 1 - \sin^2 x$ $= \int (1 - \sin^2 x) \cos x \, dx$

At this stage we need to make a substitution; we let $u = \sin x$, because sine is raised to the highest power and sine when differentiated goes to cos, which is also a part of the integral.

3) Substitution: $u = \sin x \Rightarrow \frac{du}{dx} = \cos x \Rightarrow du = \cos x \, dx$,

4) Substituting into the integral gives: $\int (1 - u^2) \, du = u - \frac{1}{3} u^3 + c$

5) Substituting back for x finally gives: $\int \cos^3 x \, dx = \sin x - \frac{1}{3} \sin^3 x + c$

Example 8

Find $\int 2x\sqrt{(2x - 5)} \, dx$ using the substitution $u = 2x - 5$

Solution

To begin, $\frac{du}{dx} = 2 \Rightarrow 2 \, dx = du$

Looking at the integral we know: (a) $2 \, dx = du$ and (b) $\sqrt{(2x - 5)} = \sqrt{u} = u^{\frac{1}{2}}$, but we still have x left to change in terms of u

When rearranged, $u = 2x - 5$, becomes $x = \frac{1}{2}(u + 5)$

Now we can make an effective substitution:

$$\int 2x\sqrt{(2x-5)}\,dx = \int \frac{1}{2}(u+5)(u)^{\frac{1}{2}}du = \frac{1}{2}\int u^{\frac{1}{2}}(u+5)du$$

$$= \frac{1}{2}\int u^{\frac{3}{2}} + 5u^{\frac{1}{2}}du = \frac{1}{2}\left(\frac{2}{5}u^{\frac{5}{2}} + \frac{10}{3}u^{\frac{3}{2}}\right) + c = \frac{1}{5}u^{\frac{5}{2}} + \frac{5}{3}u^{\frac{3}{2}} + c$$

Substituting back in terms of x finally gives: $\frac{1}{5}(2x-5)^{\frac{5}{2}} + \frac{5}{3}(2x-5)^{\frac{3}{2}} + c$

Example 9

Find $\int \frac{x}{(x^2+8)}dx$, using the substitution $u = x^2 + 8$

Solution

To begin, $\frac{du}{dx} = 2x \Rightarrow \frac{du}{2} = x\,dx$

So, $\int \frac{1.x}{x^2+8}dx = \int \frac{1}{u}\frac{du}{2} = \frac{1}{2}\int \frac{1}{u}du = \frac{1}{2}\ln|u| + c$

Substituting back in terms of x finally gives: $\frac{1}{2}\ln|x^2+8| + c$

Choosing a substitution

Sometimes the choice of substitution is left to the student. The substitution should make the integral look easier. Here are two rules of thumb:

1 Let $u =$ the expression in the brackets (see Examples 5 and 8).

2 If one part of the integral looks like the derivative of another part, then let $u =$ the bit that you can differentiate (see Examples 6,7 and 9).

6 Logarithmic Integrals

Always be on the look out for logarithmic integrals. These are of the form:

$$\boxed{\int \frac{f'(x)}{f(x)}dx = \ln|f(x)| + c}$$

The generalised formula says that if you see an integral that is a rational fraction, where the top is the differential of the bottom, then it is a logarithmic integral.

Example 10

Find: $\int \frac{8x-5}{4x^2-5x+1}dx$

Solution

Since $\frac{d}{dx}(4x^2 - 5x + 1) = 8x - 5$, which is on the top of the integral,

then answer is: $\ln|4x^2 - 5x + 1| + c$

Logarithmic integration also works when the top is a multiple of the differential of the bottom. This is shown in the following example:

Example 11

Find: $\int \frac{x + 1}{2x^2 + 4x + 7} dx$

Solution

We note that $\frac{d}{dx}(2x^2 + 4x + 7) = 4x + 4$; is 4 times the numerator.

We proceed by making the numerator what we want it to be, multiplying the integral by $\frac{1}{4}$ to make the whole thing work. Hence the numerator now becomes the differential of the denominator.

So, $\frac{1}{4}\int \frac{4x + 4}{2x^2 + 4x + 7} dx = \frac{1}{4}\ln|2x^2 + 4x + 7| + c$

Exercise: Attempt Example 9 again, by using logarithmic integration.

Logarithmic integration also applies to trigonometric expressions.

Example 12

Prove $\int \tan x \, dx = \ln|\sec x| + c$

Solution

$\int \tan x \, dx = \int \frac{\sin x}{\cos x} dx$, and since $\frac{d}{dx}(\cos x) = -\sin x$, the integral becomes

$= -\int \frac{-\sin x}{\cos x} dx = -\ln|\cos x| + c = \ln|(\cos x)^{-1}| + c = \ln\left|\frac{1}{\cos x}\right| + c$

$= \ln|\sec x| + c$, using the laws of logarithms and the fact that $\sec x = \frac{1}{\cos x}$

Integration by Parts

Integration by parts is used when we are asked to integrate a product of two functions, which bear no relation to each other, i.e. one is not a differential of the other.

We use the formula:

$$\int u\frac{dv}{dx}dx = uv - \int v\frac{du}{dx}dx$$

Your formula book is there to help you!

Integration

A common problem is which function we call u (the function to be differentiated), and which we call $\frac{dv}{dx}$ (to be integrated). This can be resolved by the following rule of thumb:

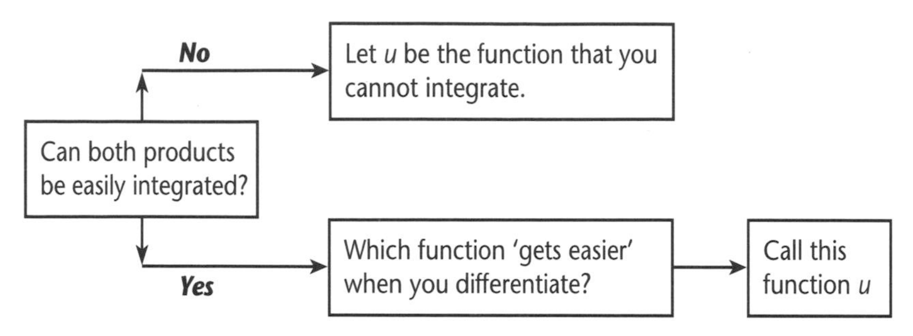

Example 13

Find $\int xe^{2x}\, dx$

Solution

We have the products x and e^{2x}. Both products can easily be integrated, but x becomes easier when you differentiate it (i.e. it becomes 1). Hence:

1) Name the functions: $u = x \qquad \frac{dv}{dx} = e^{2x}$

2) Differentiate and integrate: $\frac{du}{dx} = 1 \qquad v = \frac{1}{2}e^{2x}$

3) Apply the formula: $\int xe^{2x}dx = \frac{1}{2}xe^{2x} - \int \frac{1}{2}e^{2x}dx$

4) Integrate: $= \frac{1}{2}xe^{2x} - \frac{1}{4}e^{2x} + c$

Example 14

Find $\int x\ln x\, dx$

Solution

We have the products x and $\ln x$. The product $\ln x$ cannot easily be integrated, so it must be called u. Hence:

Differentiate and integrate: $u = \ln x, \qquad \frac{dv}{dx} = x \Rightarrow \frac{du}{dx} = \frac{1}{x}, \qquad v = \frac{1}{2}x^2$

Formula: $\int x\ln x dx = \frac{1}{2}x^2\ln x - \int \frac{1}{x}\left(\frac{1}{2}x^2\right)dx = \frac{1}{2}x^2\ln x - \frac{1}{2}\int x\, dx$

Integrate: $\frac{1}{2}x^2\ln x - \frac{1}{4}x^2 + c$

Some questions may require the repeated use of integration by parts.

Example 15

Find $\int x^2\sin x\,dx$

Solution

We have the products x^2 and $\sin x$. Since we can integrate both, we choose x^2 to differentiate because it will, when differentiated twice, be a constant.

Differentiate and integrate: $u = x^2, \frac{dv}{dx} = \sin x \Rightarrow \frac{du}{dx} = 2x, v = -\cos x$

Formula:

$\int x^2\sin x\,dx = -x^2\cos x - \int -2x\cos x\,dx = -x^2\cos x + \int \boldsymbol{2x\cos x\,dx}$ **(4)**

This time, we cannot simply integrate the part in bold italics. We must use integration by parts again to evaluate the new problem, $\int 2x\cos x\,dx$

Differentiate and integrate: $u = 2x; \frac{dv}{dx} = \cos x \Rightarrow \frac{du}{dx} = 2, v = \sin x$

Formula: $\int 2x\cos x\,dx = 2x\sin x - \int 2\sin x\,dx = 2x\sin x + 2\cos x + c$ **(5)**

Substituting **(5)** into **(4)** gives: $\int x^2\sin x\,dx = -x^2\cos x + 2x\sin x + 2\cos x + c$

8 Differential Equations

Differential Equations (DEs) are usually written in the form:

$\frac{dy}{dx}$ = (something in terms of y and x)

To solve a differential equation, you must integrate to find a relationship between y and x, with $\frac{dy}{dx}$ disappearing.

For A2 Maths, you only need to know the method of separation. This method means getting all the y terms on one side with dy on the top, and getting all the x terms on the other side with dx on top. Then you can integrate both sides, one with respect to y, the other with respect to x.

Example 16

Find the general solution of the differential equation, $(1 + x^2)\frac{dy}{dx} = xy$ in the form $y = f(x)$

Hence find the particular solution, when $x = 0, y = 5$

Solution

The general solution is the solution to the DE expressed with an unknown constant.

1) Separate the x's and y's: $\frac{1}{y}dy = \frac{x}{x^2+1}dx$

2) Put an integral sign on both sides: $\int \frac{1}{y}dy = \int \frac{x}{x^2+1}dx$

3) Logarithmic integral, so fix the RHS: $\int \frac{1}{y}dy = \frac{1}{2}\int \frac{2x}{1+x^2}dx$

4) Integrate both sides: $\ln y = \frac{1}{2}\ln(1+x^2) + c$

Before we exponentiate (take e's) of both sides, we let $c = \ln A$, where A is another constant, so $\ln A$ will also be another constant.

Hence: $\ln y = \ln(1+x^2)^{\frac{1}{2}} + \ln A$

$\Rightarrow \ln y = \ln(A(1+x^2)^{\frac{1}{2}})$

Exponentiate: $y = A(1+x^2)^{\frac{1}{2}} = A\sqrt{(1+x^2)}$ (G.S.)

The general solution is the solution of the DE, expressed with an unknown constant. The particular solution is when you solve the DE and find the constant, say A.

Use initial condition $x = 0$, $y = 5$ to find A: $5 = A\sqrt{(1+0^2)} \Rightarrow A = 5$

State the particular solution: $y = 5\sqrt{(1+x^2)}$

Example 17

Find the general solution of the differential equation, $\frac{dy}{dx} = e^{2x+y}$

Solution

This looks a nightmare, because it seems impossible to split up the x and y terms, but by the power of indices, we say $e^{2x+y} = e^{2x}e^y$, giving, $\frac{dy}{dx} = e^{2x}e^y$

Separate the variables: $\int \frac{1}{e^y}dy = \int e^{2x}dx \Rightarrow \int e^{-y}dy = \int e^{2x}dx$, since $\frac{1}{e^y} = e^{-y}$

Integrate both sides to find the general solution: $\Rightarrow -e^{-y} = \frac{1}{2}e^{2x} + c$

Volumes of Revolution

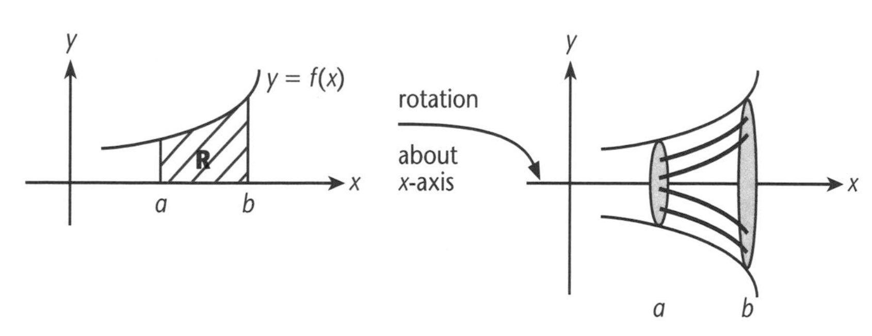

If a finite region R is rotated 360° (or $2\pi^c$) about the x-axis (as shown in the diagram) between $x = a$ and $x = b$, then the solid volume of revolution is given by the formula:

$$V = \pi \int_{x=a}^{x=b} y^2 dx \qquad\qquad V = \pi \int_{x=a}^{x=b} (f(x))^2 dx$$

Similarly, if a finite region S is rotated 360° (or $2\pi^c$) about the y-axis, between $y = a$ and $y = b$, then the solid volume of revolution is given by the formula:

$$V = \pi \int_{y=a}^{y=b} x^2 dy \qquad\qquad V = \pi \int_{y=a}^{y=b} (f(y))^2 dy$$

To use the above formulae on a curve $y = f(x)$, the finite region formed must be between the **coordinate axis** and the **curve**.

It is also important to note that when rotated through 2π radians:

- A right-angled triangular region rotates to a cone with volume: $V = \frac{1}{3}\pi r^2 h$ See example 19
- A rectangular region rotates to a cylinder with volume: $V = \pi r^2 h$

Example 18

Find, to 3 significant figures, the volume of revolution formed when the region bounded by the curve $y = e^x$, the lines $x = 1$, $x = 3$ and the x-axis, is rotated 360°, about the x-axis.

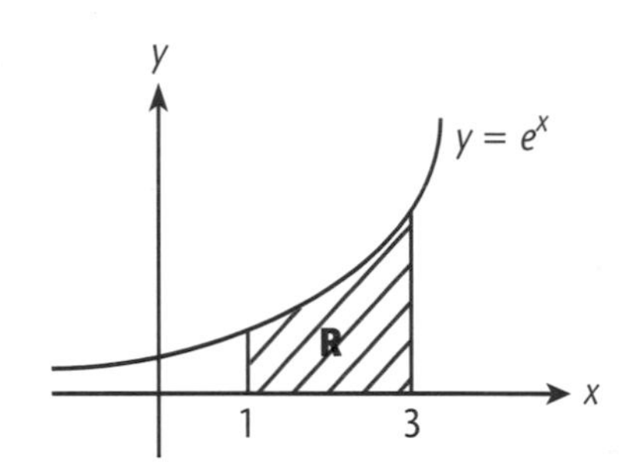

Solution

Apply the x-axis formula: $V = \pi \int_1^3 y^2 dx$

Substitute for y: $V = \pi \int_1^3 (e^x)^2 dx = \pi \int_1^3 e^{2x} dx$

Evaluate: $V = \pi \left[\frac{e^{2x}}{2}\right]_1^3 = \pi \left[\frac{e^6}{2} - \frac{e^2}{2}\right] = 622.098... = 622 \text{ (3sf)}$

Example 19

Calculate the volume of revolution, in terms of π, when the finite region R, as shown in the diagram opposite, is rotated $2\pi^c$ about the y-axis.

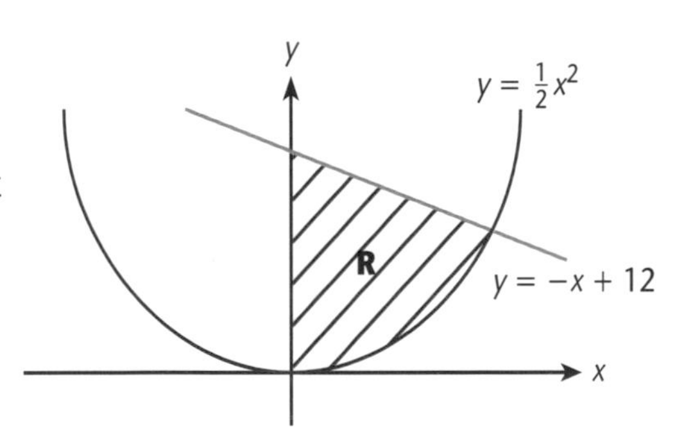

Solution

Curve = line $\Rightarrow \frac{x^2}{2} = -x + 12 \Rightarrow x^2 = -2x + 24 \Rightarrow x^2 + 2x - 24 = 0$

$\Rightarrow (x + 6)(x - 4) = 0 \Rightarrow x = -6$ (reject!) and $x = 4$ (accept!)

$\Rightarrow$ When $x = 4$, $y = \frac{4^2}{2} = 8$. Hence curve = line at (4, 8)

When $x = 0$, line $\Rightarrow y = -0 + 12 = 12 \Rightarrow$ line cuts y-axis at (0, 12)

The region R is split into two regions; R_1 and R_2.

Region R_1 is between the y-axis and the curve:
$\Rightarrow$ Use the y-axis formula

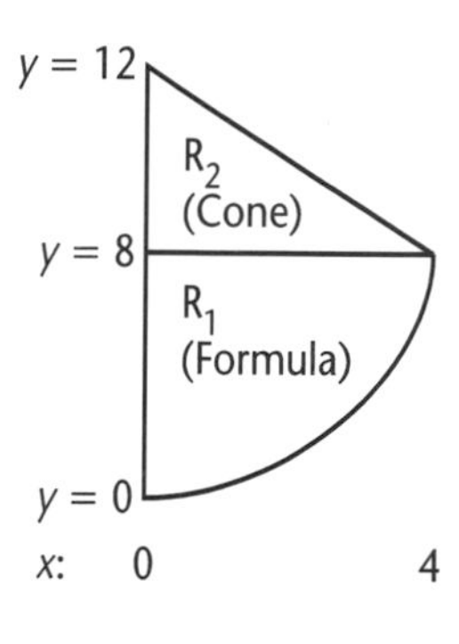

$$V_{R1} = \pi \int_0^8 x^2 dy = \pi \int_0^8 2y\, dy \text{ (since curve } x^2 = 2y\text{)}$$

$$= \pi \left[y^2 \right]_0^8 = \pi (64 - 0) = 64\pi$$

The region R_2 is a right-angled triangular region, which rotates to a cone.

Cone: $r = 4 - 0 = 4$, $h = 12 - 8 = 4 \Rightarrow V_{R2} = \frac{1}{3}\pi(4)^2 4 = \frac{64\pi}{3}$

Total volume: $V_R = V_{R1} + V_{R2} = 64\pi + \frac{64\pi}{3} = \frac{256\pi}{3}$ (units)^3

Integration

40 minutes

Use your knowledge

1 a) Given that $f(x) = \dfrac{8x + 7}{(2x^2 + 7)(4 - x)}$ can be expressed in form $\dfrac{Ax + B}{2x^2 + 7} + \dfrac{C}{4 - x}$, find the constants A, B and C.

This is a partial fraction question. See the algebra chapter for help!

b) Hence evaluate $\displaystyle\int_1^3 f(x)dx$, leaving your answer in the form $\ln q$, where q is an integer.

Integrate the partials! Both terms are ln integrals. Remember that ln 1 = 0

2 a) Find $\int x\cos 3x \, dx$

By parts. Let u = x, etc

b) Use the substitution, $x = 3\sin u$, to rewrite $\displaystyle\int_0^{\frac{3}{2}} \sqrt{(9 - x^2)}dx$ in the form $k\displaystyle\int_0^{a} \cos^2 u du$, where k and a are constants to be found.

Use the identity: $\cos^2 x + \sin^2 x \equiv 1$ Remember to change the limits! Work in radians

c) Hence evaluate $\displaystyle\int_0^{\frac{3}{2}} \sqrt{(9 - x^2)}dx$, leaving your answer in the form $e\sqrt{3} + f\pi$, where e and f are rational constants, which need to be found.

Use the cos 2x identity on the 'u' version of the integral

3 a) Use the method of integration by parts to find $\int we^{-w}dw$

Choose: u = w

b) Using the substitution $x = e^w$, show that:

$$\int \frac{3 - \ln x}{x^2} \text{ becomes } \int \frac{3 - w}{e^w} dw$$

Use $x^2 = e^w e^w$, to help you!

c) Hence, find $\displaystyle\int \frac{3 - \ln x}{x^2} dx$

Use the w version of the integral. Split up the terms. Use part (a)

Test your knowledge

The functions f and g are defined by:

$f(x) = x^2 + 10, x \in \mathbb{R}$ and $g(x) = 3x - 2, x \in \mathbb{R}$

a) Sketch the graphs of the two functions f and g, showing where they cut the coordinate axis.

b) Hence state the range of the functions f and g.

c) Explain whether or not (i) f, (ii) g have inverse functions, and if so find the inverse.

d) Solve the equation $f(x) = gg(x)$.

The function f is defined by $f(x) = 6x - x^2, x \in \mathbb{R}$

a) Sketch the graph of the function f, showing where it cuts the coordinate axis.

b) Hence explain why $f(x)$ has no inverse.

The function h is defined by $h(x) = 6x - x^2, x \geq c$, is a one–one function.

c) Find the minimum value of c.

d) Find the inverse function, $h^{-1}(x)$

3 The graph of $y = f(x)$, as sketched opposite, is such that $f(x) = 5$ for $x \leq 1$ and $x \geq 6$. Sketch on separate axis the graphs with equations:

a) $y = f(2x) - 4$ b) $y = \frac{1}{2}f(x + 1)$

c) $y = -f(-x)$

stating the new coordinates of the corresponding points A, B and C.

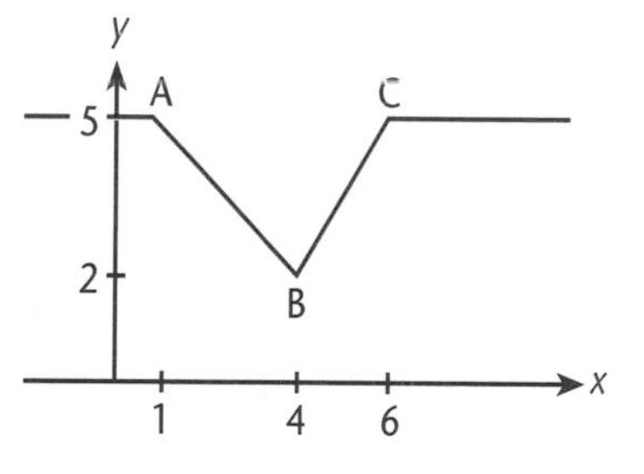

Solve the following equations: a) $|3x - 2| = 10$ b) $|1 - 4x| = |2x + 5|$

Answers

1 a) $f(x)$: a parabola cutting the y-axis at (0, 10) $g(x)$: a straight line cutting the coordinate axis at (0, −2) and ($\frac{2}{3}$, 0) **b)** f: $f(x) \geq 10$ or $y \geq 10$ g: $g(x) \in \mathbb{R}$ or $y \in \mathbb{R}$ **c) i)** f is a many–one mapping $\Rightarrow f^{-1}$ is a one–many mapping and is not a function. **ii)** g is a one–one mapping $\Rightarrow g^{-1}$ is also a one–one mapping, and hence a function. $g^{-1}(x) = \frac{x + 2}{3}, x \in \mathbb{R}$ **d)** 3, 6

2 a) An inverted parabola cutting the coordinate axis at (0,0) and (6, 0) **b)** $f(x)$ is a many–one mapping $\Rightarrow f^{-1}(x)$ would be a one–many mapping, which is not a function. **c)** 3 **d)** $h^{-1}(x) = \sqrt{(9 - x)} + 3$

3 a) A(0.5, 1) , B(2, −2) , C(3, 1) **b)** A(0, 2.5), B(3, 1), C(5, 2.5) **c)** A(−1, −5), B(−4, −2), C(−6, −5) **4 a)** 4, $\frac{-8}{3}$ **b)** 3, $\frac{-2}{3}$

If you got them all right, skip to page 61

Functions

Improve your knowledge

What is a Function?

A function is a process which maps (i.e. converts) 'input' values to 'output' values. A function consists of three components: an *input*, *rule* and *output*.

Example 1

$f(x) = x^2$ is a function.

The input is a value, say x. The rule of this function is to square the input.

Hence the output is x^2.

The domain is the values that can be put into a function. The range is the values that can be output from a function.

When representing a function as a graph, the domain is the x-values the function can take and the range is the y-values.

A function is valid, when for every input, there is only **one** output.

f

x → Rule: Square it → x^2

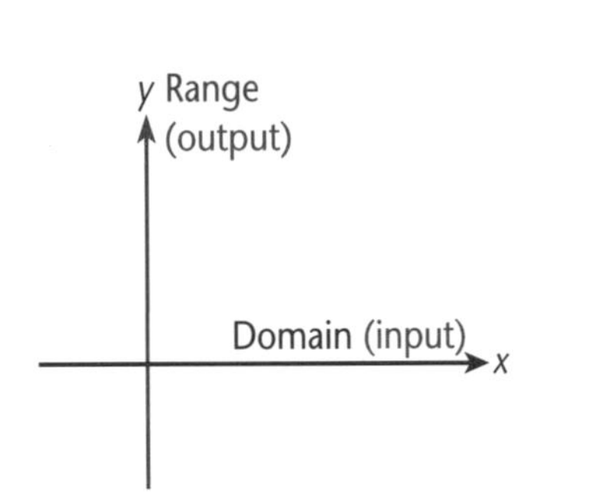

Example 2

$f(x) = \pm\sqrt{x}$ is not a function because it has more than one output.

Eg. if $x = 9$ (input), then $f(9) = \pm 3$, gives two outputs, which is not possible for a function.

Domain and Range

The domain of a function is usually given to you in an examination question. For a function, say $f(x) = x^2 + 2$, there are many possible choices of domain, each giving rise to a range.

Example 3

Give the range of the following functions:

a) $f(x) = x^2 + 2\ ,\ x \in \mathbb{R}$ b) $g(x) = x^2 + 2\ ,\ x \geq 0$ c) $h(x) = x^2 + 2,\ x \leq -1$

Solution

To do this we will sketch the graphs of each of the functions. When sketching the graphs it is important to look at the domain, because this tells us what x-values we can only draw the graph for. $f(x) = x^2 + 2$ is the $y = x^2$ graph moved up two units in the y-direction.

Functions

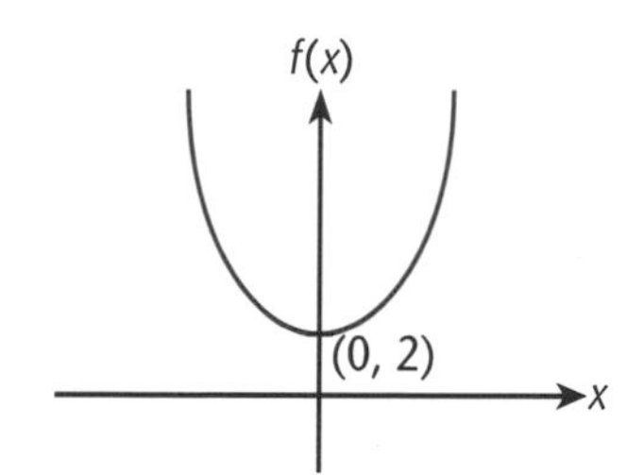

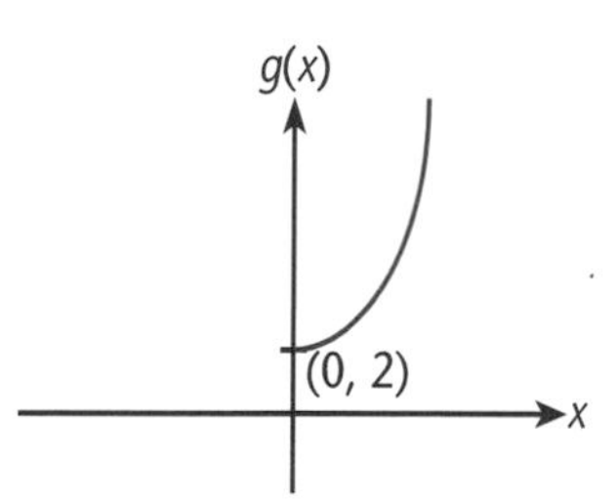

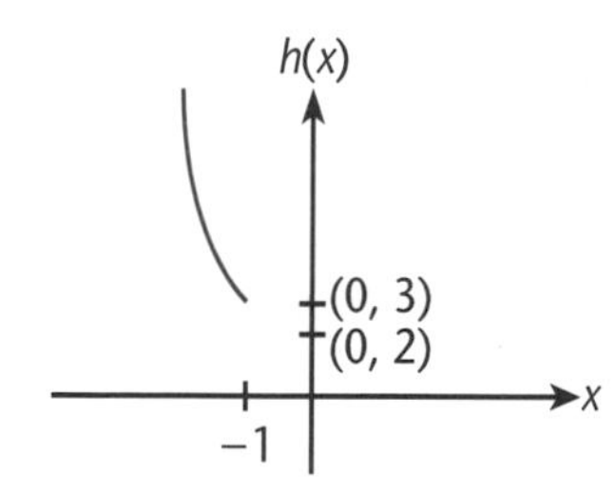

a) Range:
   $f(x) \geq 2$ or $y \geq 2$

b) Range
   $g(x) \geq 2$ or $y \geq 2$

c) Range
   $h(x) \geq 3$ or $y \geq 3$

Sometimes it is necessary to restrict the domain so that a function 'makes sense'.

- $f(x) = \frac{1}{x}$ is not defined at $x = 0$, so $x \neq 0$
- $g(x) = \sqrt{x}$ is not possible for negative x values, so $x \geq 0$

Composite Functions

When two or more functions are combined together to form a new function the result is called a composite function. With a composite function we appear to work backwards. For $fg(x)$, first g is applied to x, followed by f.

Example 4

For the functions: $f(x) = x + 3$, $x \subset \mathbb{R}$, and $g(x) = x^2 - 1$, $x \in \mathbb{R}$, find:

a) $fg(x)$ b) $gf(x)$ and   c) $ffg(x)$

Solution

a) Work backwards from x: $fg(x) = f(x^2 - 1)$
   Put g into f. Every time we see x in the f function we replace it by what's in g, i.e. $x^2 - 1$.
   This gives $fg(x) = x^2 - 1 + 3 = x^2 + 2$, $x \in \mathbb{R}$

b) $gf(x) \Rightarrow f$ goes into $g \Rightarrow$ replace x in g by $x + 3$
   This gives $gf(x) = g(x + 3) = (x + 3)^2 - 1$, $x \in \mathbb{R}$

c) $ffg(x) = ff(x^2 - 1) = f(x^2 - 1 + 3) = f(x^2 + 2) = x^2 + 2 + 3 = x^2 + 5$, $x \in \mathbb{R}$

Since $fg(x)$ is a function, it will have a domain. The domain will usually be the domain of the function that x is applied to first, in this case, g. Since the domain of g is $x \in \mathbb{R}$, the domain of $fg(x)$ is also $x \in \mathbb{R}$

Sometimes we must restrict the domain of the first function to make the composite function work.

Example 5

For the functions: $g(x) = \frac{1}{x}$, $x \in \mathbb{R}$, $x \neq 0$ and $h(x) = x + 5$, $x \in \mathbb{R}$, $x \neq -5$

a) find $gh(x)$, stating its domain, and
b) solve the equation $gh(x) = h(x)$

Solution

a) h goes into $g \Rightarrow gh(x) = g(x + 5) = \frac{1}{x+5}$. The domain of $gh(x)$ is $x \in \mathbb{R}$, $x \neq -5$, to prevent the function being undefined at $x = -5$. This means that $x \neq -5$ must be a part of the original function $h(x)$

b) $\frac{1}{x+5} = x + 5 \Rightarrow 1 = (x + 5)(x + 5) \Rightarrow 1 = x^2 + 10x + 25$
$\Rightarrow x^2 + 10x + 24 = 0$

Factorising gives: $(x + 6)(x + 4) = 0 \Rightarrow x = -6, -4$

One–One and Many–One Functions

A **many–one** function, $f(x)$, is a function that for *some* values of $f(x)$ there exists *more than one* corresponding values for x.

A **one–one** function, $f(x)$, is a function that for *every* value of $f(x)$ there exists *only one* corresponding value of x.

- $f(x) = x^2$, $x \in \mathbb{R}$, is a many–one function, because when $f(x) = 9$, x could be either 3 or -3.
- $g(x) = 2x + 3$, $x \in \mathbb{R}$, is a one–one function because every $g(x)$ value has one corresponding x value.

We can spot whether a function is one–one by drawing its graph. We know that each $f(x)$ value must correspond to only one x-value, so if we draw horizontal lines anywhere on the graph, they will only cut the graph *at most*, once. If any horizontal line cuts the graph more than once then it is a many–one.

$f(x) = x^3$, $x \in \mathbb{R}$

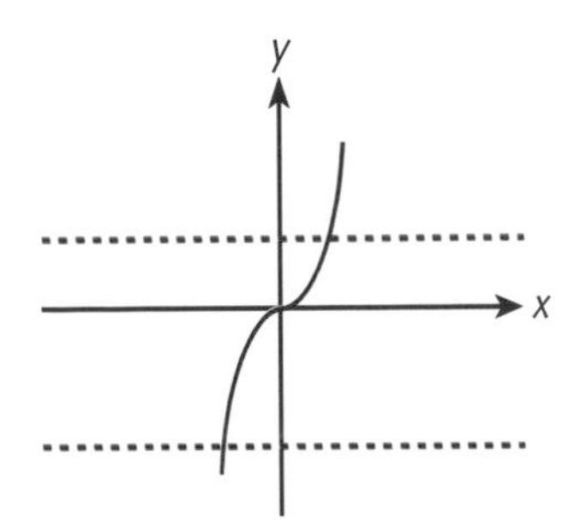

One–one $\Rightarrow f^{-1}(x)$ exists

$f(x) = (x - 1)(x + 2)(x + 1)$, $x \in \mathbb{R}$

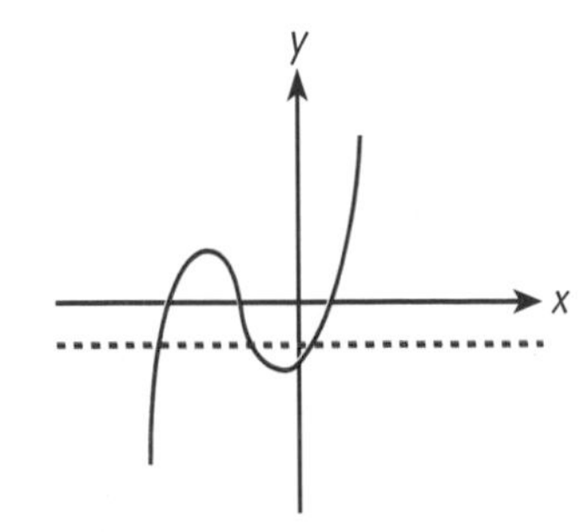

Many–one $\Rightarrow f^{-1}(x)$ does not exist!

Functions

Inverse Functions

An inverse function $f^{-1}(x)$ is a function that 'undoes' what $f(x)$ has done. The function $f(x)$ must be a one–one function to have an inverse. The inverse of a many–one function does not exist. This is because the inverse would be a one–many mapping, which has more than one output, and thereby is not a function.

For inverse functions you need to know that:

- the domain of a function $f(x)$ becomes the range of the inverse function $f^{-1}(x)$
- the range of a function $f(x)$ becomes the domain of the inverse function $f^{-1}(x)$
- the graph of $f^{-1}(x)$ is a reflection of the graph of $f(x)$ in the line $y = x$

Example 6

For the function $f(x) = 7x - 3$, $1 \leq x \leq 5$, find

a) the range, b) the inverse function $f^{-1}(x)$, stating its domain and range.

Solution

a) The function is a straight line with positive gradient starting from $x = 1$ and ending at $x = 5$. The smallest value is $f(1) = 7 - 3 = 4$; the largest value is $f(5) = 35 - 3 = 32$.

Hence the range of the function is $4 \leq f(x) \leq 32$

b) Write down $y = f(x)$: $y = 7x - 3$

- Swap the x's and y's: $x = 7y - 3$
- Rearrange to make y the subject: $x + 3 = 7y \Rightarrow \frac{x+3}{7} = y$
- Then write $f^{-1}(x) = y$: $f^{-1}(x) = \frac{x+3}{7}$

The domain of $f^{-1}(x)$ is the range of $f(x) \Rightarrow$ Domain of $f^{-1}(x)$ is $4 \leq x \leq 32$.
The range of $f^{-1}(x)$ is the domain of $f(x) \Rightarrow$ Range of $f^{-1}(x)$ is $1 \leq f^{-1}(x) \leq 5$.

Example 7

For the function f, defined for $x \in \mathbb{R}$, by $f: x \rightarrow e^x + 2$, find:

a) the inverse function $f^{-1}(x)$, b) the range of $f^{-1}(x)$.

Hence on the same coordinate axis sketch the graph of $y = f(x)$ and its inverse.

Solution

a) $y = e^x + 2$, so write $x = e^y + 2$
$x - 2 = e^y \Rightarrow \ln(x - 2) = y$
Finally $f^{-1}(x) = \ln(x - 2)$

b) Since the domain of $f(x)$ is $x \in \mathbb{R}$, then the range of the inverse $f^{-1}(x)$ is: $y \in \mathbb{R}$

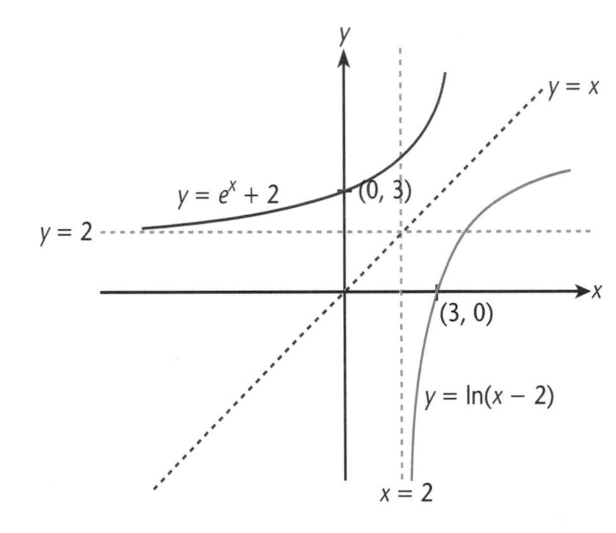

Curve of e^x is moved up 2 units.

$f(x)$ has: domain $x \in \mathbb{R}$
range $y > 2$

$f^{-1}(x)$ has: domain $x > 2$
range $y \in \mathbb{R}$

$y = \ln(x - 2)$ is found by reflecting $y = e^x + 2$ in the line $y = x$

Trying to Find Inverses of Many–One Functions

We cannot find the inverse of a many–one function. However, by *restricting the domain*, we can turn a many–one function into a one–one function, which has an inverse. For quadratic functions, we usually cut the function in half and disregard one half.

$f(x) = x^2, x \in \mathbb{R}$

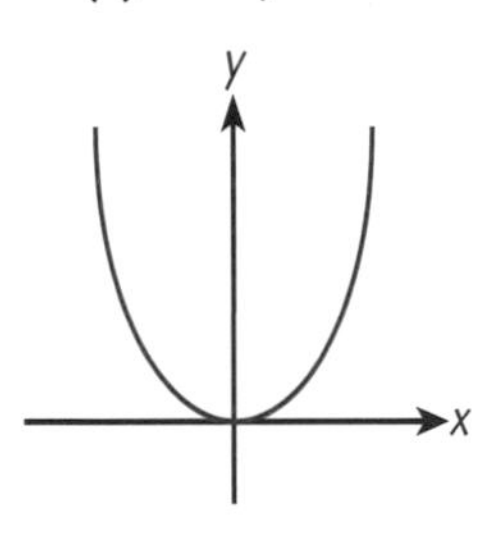

Many–one
No inverse

$f(x) = x^2, x \geq 0$

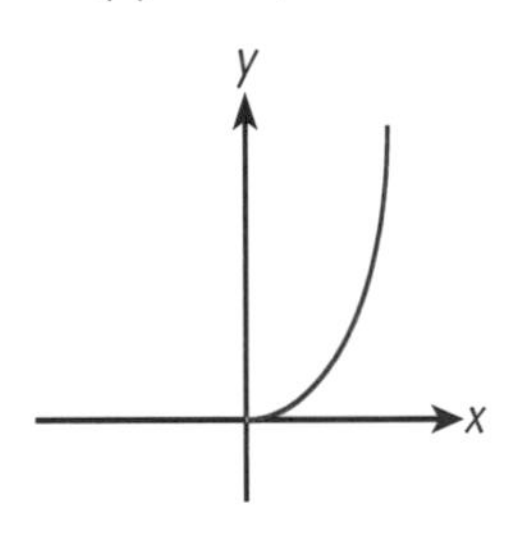

One–one
$f^{-1}(x) = \sqrt{x}, x \geq 0$

$f(x) = x^2, x \leq 0$

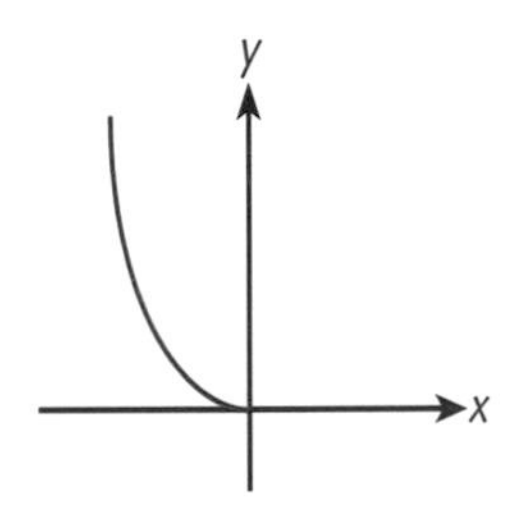

One–one
$f^{-1}(x) = -\sqrt{x}, x \geq 0$

Remember: For a quadratic function, the x-coordinate of the maximum (or minimum) on the curve is found by averaging the two roots.

Example 8

The function f is defined by $f(x) = x^2 - 10x, x \in \mathbb{R}$

a) Sketch the function

b) Hence explain why $f(x)$ has no inverse

The function g defined by $g(x) = x^2 - 10x$, $x \geq k$, is a one–one function.

c) Find the minimum value of k, d) Find $g^{-1}(x)$

Solution

a) $y = x^2 - 10x = x(x - 10) = 0$
$\Rightarrow x = 0$ and $x = 10$ are roots.

This is a quadratic, with minimum at $x = \frac{0 + 10}{2} = 5$

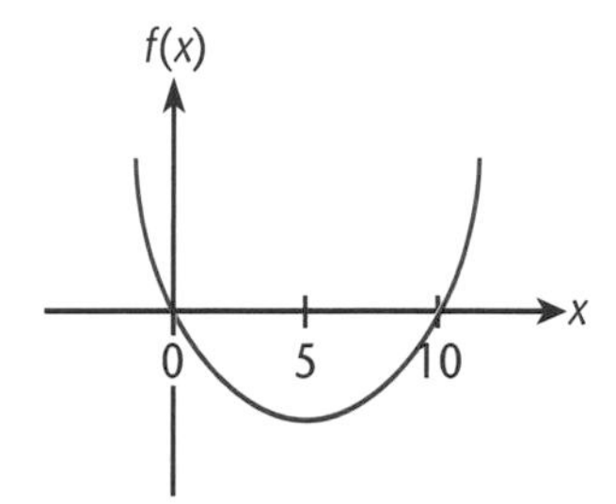

b) The function is a many–one mapping. Hence, the inverse is a one–many mapping, which is not a function.

c) $g(x)$ looks like $f(x)$ except the domain of $g(x)$ is different. If we cut out part of the graph of $f(x)$ where $x < 5$, then the function will be one–one and will have an inverse. Hence the function will be valid for $x \geq 5$. So the minimum value $k = 5$.

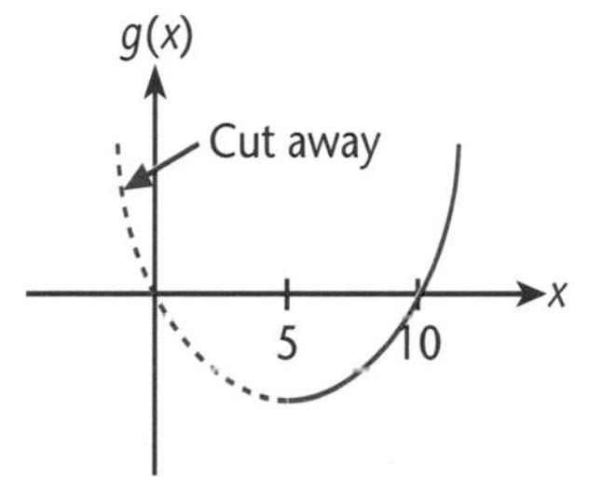

d) We need to use the process of completing the square to help us to find the inverse.

$y = x^2 - 10x$, so for the inverse $x = y^2 - 10y$
$x = (y - 5)^2 - 25 \Rightarrow x + 25 = (y - 5)^2 \Rightarrow \sqrt{x + 25} = y - 5$
$\Rightarrow \sqrt{x + 25} + 5 = y$
Finally $g^{-1}(x) = \sqrt{x + 25} + 5$, $x \geq -25$

Types of Functions

You need to be aware of the following types of functions:

- An even function is a function which reflects onto itself through the y-axis. It is defined by $f(x) = f(-x)$. Examples of even functions are $y =: x^2, x^4, \cos x, |x|$.
- An odd function is a function which is unchanged after rotation of 180° about the origin (or is unchanged after a reflection in the x-axis followed by a reflection in the y-axis.) It is defined by $f(-x) = -f(x)$. Examples of odd functions are $y =: x, x^3, \sin x, \tan x$.

- A periodic function is a function that repeats itself at regular intervals of x. It is defined by $f(x) = f(x + nT)$, where $n = 1, 2, 3,....$ and T is the period. Examples of periodic functions are $y =$: $\sin x$ (period of 2π), $\tan x$ (period of π).

3 Transformations of Curves

You need to be able to learn the basic graphs like $y =$: x^2, x^3, $\frac{1}{x}$, $\frac{1}{x^2}$, $\sin x$, $\cos x$, $\tan x$, $|x|$, e^x, $\ln x$, etc. Once you have learnt how to draw these graphs, you can use the method of transformations of curves to draw more complicated curves.

If you start with the basic graph of $y = f(x)$, then:

- $y = f(x) + A$ graph moves up A units
- $y = f(x) - A$ graph moves down A units
- $y = f(x - A)$ graph moves to the right A units
- $y = f(x + A)$ graph moves to the left A units
- $y = f(Ax)$ stretch scale factor (SF) $\frac{1}{A}$ parallel to the x-axis
- $y = Af(x)$ stretch scale factor A parallel to the y-axis
- $y = -f(x)$ reflection of the graph in the x-axis
- $y = f(-x)$ reflection of the graph in the y-axis

where A is a positive real number.

Example 9

Sketch the graphs of the following:

a) $y = (x + 3)^2$ b) $y = (x - 2)^2 + 3$ c) $y = 5 - (x + 2)^2$

Solution

a)

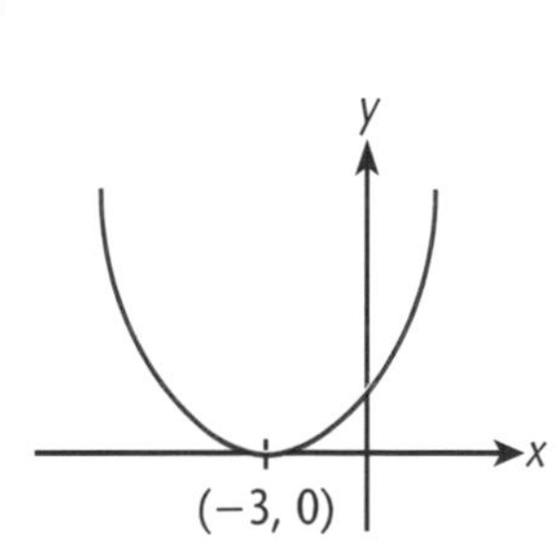

b)

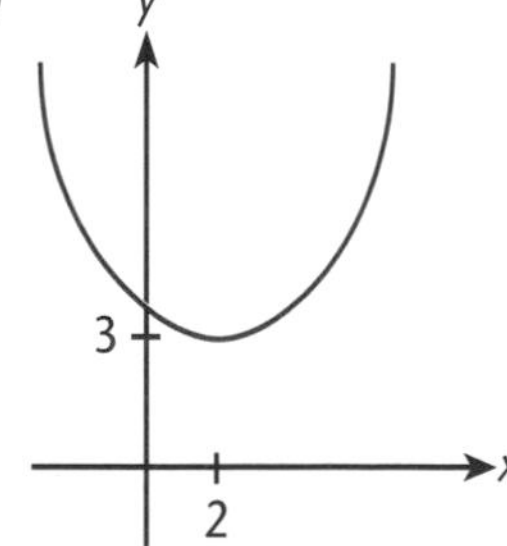

c)

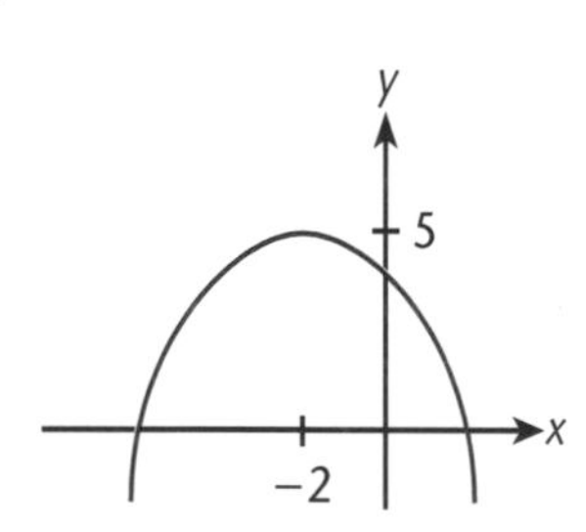

All three of the above graphs are based on the transformation of the curve $y = x^2$

Graph a): $y = x^2$ is moved three units to the left
Graph b): $y = x^2$ moves two units to the right and three units up
Graph c): $y = x^2$ moves two units to the left; reflects through the x-axis; and then moves up by 5 units

Example 10

The diagram opposite shows the graph of $y = f(x)$. Sketch on separate axis the graphs with equations:

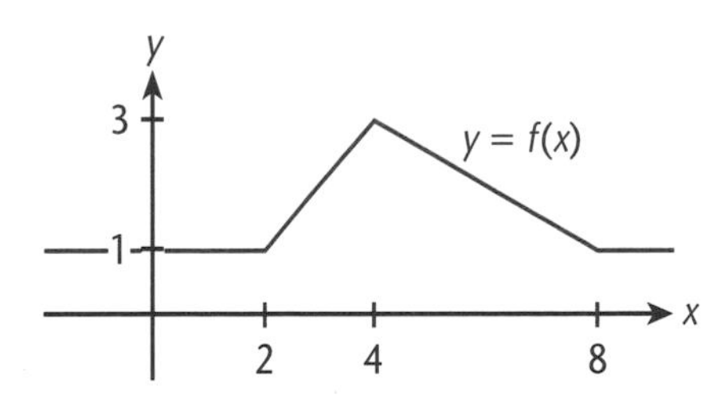

a) $y = f(2x) - 2$
b) $y = \frac{1}{2}f(x + 3)$

Solution

a)

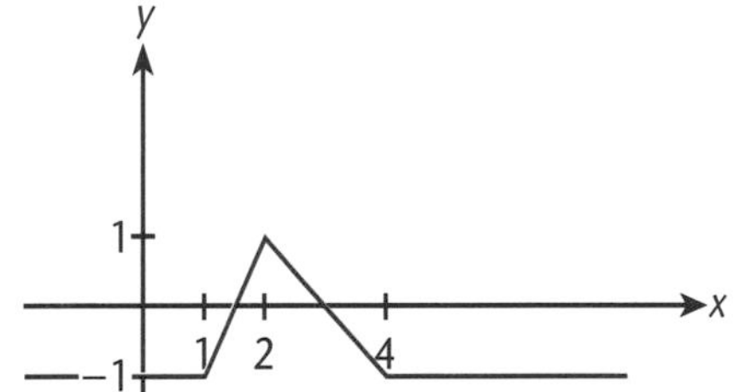

b)

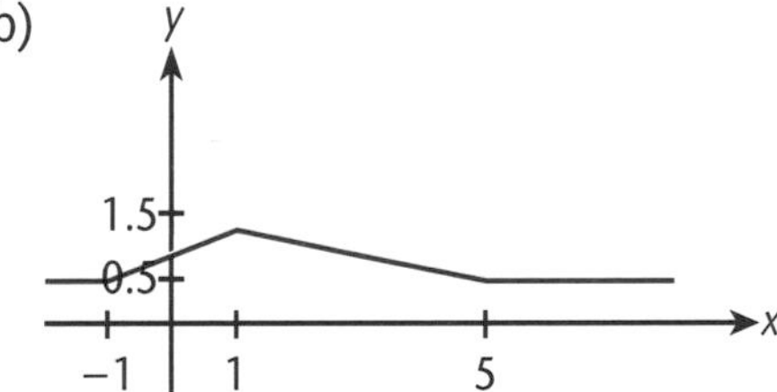

a) $y = f(x)$ stretches SF $\frac{1}{2}$ in the x-direction; then moves down 2 units.

b) $y = f(x)$ moves to the left 3 units; then stretches SF $\frac{1}{2}$ in the y-direction.

4 The Modulus Function

The modulus of x, (written $|x|$), gives us the positive value of x.
For example, $|3| = 3$ and $|-5| = 5$

To solve an equation of the form $|ax + b| = c$, we let the inside of the modulus equal to plus or minus the value of the constant c.

To solve an equation of the form $|ax + b| = |cx + d|$, we 'square both sides' and then solve.

In other circumstances we need to sketch a graph to help us.

Example 11

Solve the equations: a) $|2x + 1| = 7$ b) $|4x + 1| = |2x + 5|$

Solution

a) $2x + 1 = 7 \Rightarrow 2x = 6 \Rightarrow x = 3$
$2x + 1 = -7 \Rightarrow 2x = -8 \Rightarrow x = -4$
Solutions are $x = 3, -4$

b) $(4x + 1)^2 = (2x + 5)^2 \Rightarrow 16x^2 + 8x + 1 = 4x^2 + 20x + 25$
$\Rightarrow 12x^2 - 12x - 24 = 0 \Rightarrow 12(x^2 - x - 2) = 0 \Rightarrow 12(x - 2)(x + 1) = 0$
Solutions are $x = 2, -1$

To draw the graph of $y = |\textit{something}|$, you draw the graph without the modulus, and reflect the negative y-values (in blue) through the x-axis, as shown in the examples below.

Example 12

Sketch the graphs of: a) $y = |x - 2|$ and b) $y = |x^2 - 10x|$

Solution

a)

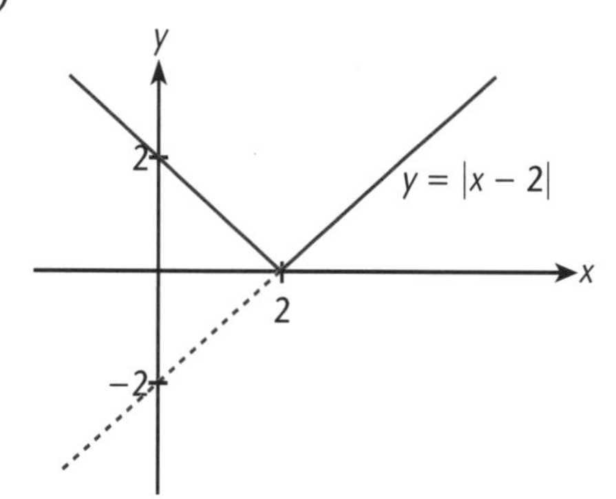

b)

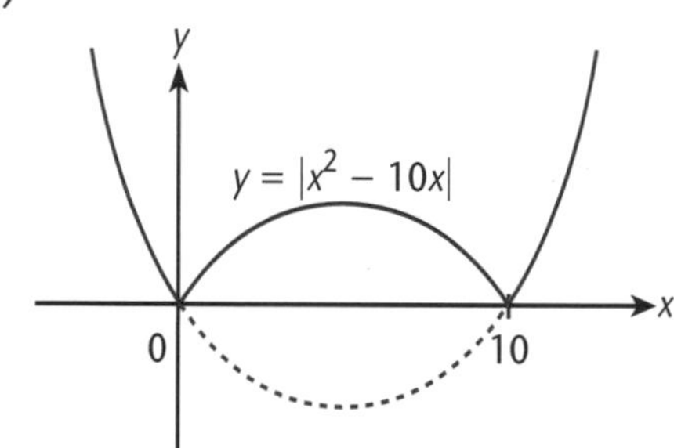

(Refer to graph in Example 8)

Use your knowledge

a) Describe the successive graph transformations that map the graph of $y = \frac{1}{x}$ to the graph $y = a + \frac{1}{(x + b)}$, where a and b are positive constants.

What does $\frac{1}{x + b}$ do? Then what does the '+a' mean?

b) Sketch the graph of $y = a + \frac{1}{(x + b)}$, giving the equations of the asymptotes.

Apply the successive transformations found in (a) to the graph of $y = \frac{1}{x}$

The functions f and g are defined as:
$f(x) = \frac{1}{x + 2} + 1,\ x \geq 0$ and $g(x) = \ln x,\ x > 0$

c) Sketch the graph of the function $f(x)$, showing where it cuts the coordinate axis.

Look at part (b) for help. Remember f(x) is only defined for $x \geq 0$

d) Hence state the range of $f(x)$

Look at your graph in (c)! What y values is it defined?

The inverse function of $f(x)$ is called $f^{-1}(x)$

e) State the domain and range of $f^{-1}(x)$

Domain $f^{-1} \equiv$ Range f. Range f^{-1} = Domain f

f) Find the value of x, to 2 decimal places, satisfying the equation $f(x) = g(e^{2x})$

ln and e cancel out. You need to solve a quadratic by the formula

The function g is defined as $g(x) = \frac{x - 1}{x + 4},\ x \in \mathbb{R},\ x \neq -4$

Put 'g' into 'g'.

a) Express $gg(x)$ in the form $\frac{a}{x + b}$, where a and b are constants to be determined.

Replace 'x' in g by $\frac{x - 1}{x + 4}$. Watch out for nasty algebra!!

b) Find the inverse function $g^{-1}(x)$.

Change the letters around and make y the subject

Vectors

Test your knowledge

1 The position vectors of the points A, B and C are $6\mathbf{i} + 2\mathbf{j}$, $3\mathbf{i} + 8\mathbf{j} + \mathbf{k}$ and $-2\mathbf{i} + 5\mathbf{j} - \mathbf{k}$ respectively. Find

a) $\overrightarrow{AB}$

b) The distance AC.

2 a) i) Find the scalar product of $\mathbf{i} - \mathbf{j} + 2\mathbf{k}$ and $2\mathbf{i} - \mathbf{k}$.

ii) State what you can conclude about these two vectors

b) Use vector methods to find angle BAC from question 1.

3 This question refers to the points A, B and C from question 1.

a) Find a vector equation of the line AB

b) Point D has position vector $5\mathbf{i} + 6\mathbf{j} + \mathbf{k}$. Find whether point D is on the line AB

c) Point E has position vector $4\mathbf{i} + 8\mathbf{j} + \mathbf{k}$

i) Find the equation of the line CE

ii) Determine whether lines CE and AB intersect

iii) Find the angle between lines CE and AB.

4 The point P is on the line with vector equation $\mathbf{r} = 6\mathbf{i} + 2\mathbf{j} + \mathbf{k} + \lambda(\mathbf{i} - 2\mathbf{j} + \mathbf{k})$. The point Q has position vector $6\mathbf{i} - 2\mathbf{j}$.

a) Determine the possible position vector(s) of P, given that angle POQ = 90°

b) Determine the shortest distance from the line to the origin.

Answers

1 a) $-3\mathbf{i} + 6\mathbf{j} + \mathbf{k}$ **b)** $\sqrt{74}$

2 a) i) 0 **ii)** they are perpendicular **b)** 45.4°

3 a) $\mathbf{r} = 6\mathbf{i} + 2\mathbf{j} + \lambda(-3\mathbf{i} + 6\mathbf{j} + \mathbf{k})$ (or $\mathbf{r} = 3\mathbf{i} + 8\mathbf{j} + \mathbf{k} + \lambda(-3\mathbf{i} + 6\mathbf{j} + \mathbf{k})$)

b) It is not

c) i) $\mathbf{r} = -2\mathbf{i} + 5\mathbf{j} - \mathbf{k} + \mu(6\mathbf{i} + 3\mathbf{j} + 2\mathbf{k})$

(or $\mathbf{r} = 4\mathbf{i} + 8\mathbf{j} + \mathbf{k} + \mu(6\mathbf{i} + 3\mathbf{j} + 2\mathbf{k})$)

ii) They do not **iii)** 87.6°

4 a) $2.8\mathbf{i} + 8.4\mathbf{j} - 2.2\mathbf{k}$ **b)** $\frac{1}{2}\sqrt{158}$

✔ If you got them all right, skip to page 69

Improve your knowledge

1

- Vectors can be written as columns, like $\begin{pmatrix} 2 \\ -3 \\ 5 \end{pmatrix}$ or using **i**, **j** and **k**, like $2\mathbf{i} - 3\mathbf{j} + 5\mathbf{k}$

  You can use whichever you find easier, but it's best not to mix them! We'll use **i**, **j** and **k** here – to convert this sort of vector to a column vector, put the number of **i**'s in the top of the column, the number of **j**'s in the middle, and the number of **k**'s in the bottom of the column.
- A position vector tells you where a point is relative to the origin. It works in the same way as coordinates – a point with position vector $2\mathbf{i} - 3\mathbf{j} + 5\mathbf{k}$ has coordinates $(2, -3, 5)$
- If you know that point A has position vector **a** and point B has position vector **b**, then the vector $\overrightarrow{AB} = \mathbf{b} - \mathbf{a}$
- To find the magnitude or length of a vector, square the number of **i**'s, the number of **j**'s and the number of **k**'s, add the answers together, then square root (like Pythagoras). NB: You do NOT square the actual **i**, **j** or **k**!
  e.g. $\mathbf{v} = 2\mathbf{i} - 3\mathbf{j} + 5\mathbf{k}$. Magnitude of $\mathbf{v} = |\mathbf{v}| = \sqrt{(2^2 + (-3)^2 + 5^2)} = \sqrt{38}$
- The distance between points A and B is the magnitude of the vector $\overrightarrow{AB}$

2

The Scalar Product and its Use

The scalar (or dot) product of two vectors is given by:

$\mathbf{a}.\mathbf{b} = |\mathbf{a}|\ |\mathbf{b}| \cos\theta$ where θ is the angle between the vectors.

You MUST use a dot, not a × sign!

Special case: if **a** and **b** are at right angles, then $\theta = 90°$, so $\mathbf{a}.\mathbf{b} = 0$

Also, you find the dot product of two vectors by multiplying together their **i** components, multiplying together their **j** components, multiplying together their **k** components then adding up.

This can be used to find the angle between two vectors.

Example 1

$\mathbf{a} = 3\mathbf{i} - 4\mathbf{j} + \mathbf{k}$ $\quad$ $\mathbf{b} = -2\mathbf{i} + 5\mathbf{j} + 2\mathbf{k}$

Find the angle between **a** and **b**

Solution

Step 1 Find **a.b**

$\mathbf{a.b} = 3 \times -2 + -4 \times 5 + 1 \times 2$
$= -24$

Keep everything in surd form until the end

Step 2 Find $|\mathbf{a}|$, $|\mathbf{b}|$ and use the formula to find $\cos\theta$, and hence θ

$|\mathbf{a}| = \sqrt{(3^2 + (-4)^2 + 1^2)} = \sqrt{26}$
$|\mathbf{b}| = \sqrt{((-2)^2 + 5^2 + 2^2)} = \sqrt{33}$
So $-24 = \sqrt{26}\sqrt{33}\cos\theta$
$\theta = 145°$

The dot product can also be used to find an angle in a triangle (or other shape).

You find the angle between the vectors corresponding to the sides you need the angle between. However, you have to be careful that you choose

either both vectors pointing away from the angle

or both vectors pointing towards the angle

This is very important!

For example, to find the angle ABC shown below, we could either use $\overrightarrow{BA}$ and $\overrightarrow{BC}$, or $\overrightarrow{AB}$ and $\overrightarrow{CB}$.

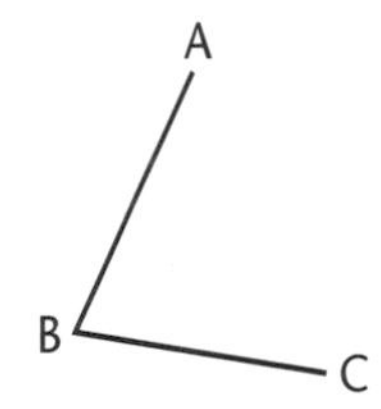

3 Vector Equations of Lines

The vector equation of a line looks like:

$$\mathbf{r} = \mathbf{a} + \lambda\mathbf{d}$$

- **r**: the position vector of any point on the line – different values of λ give you the different points
- **a**: the position vector of a point you know is on the line
- λ: a parameter – which is a number that can take any value
- **d**: the direction vector of the line – it tells you the direction the line is going in

In an actual line equation, **a** and **d** will be actual vectors, but **r** and λ will not be replaced by anything.

Finding the Equation of a Line

To find the vector equation of a line, you need to have either

- a point it goes through and its direction, or
- two points it goes through.

Example 2

a) Line L passes through point A, which has coordinates (5, 3, −2), and is parallel to the vector $4\mathbf{i} - 2\mathbf{j} - 3\mathbf{k}$. Find the vector equation of line L

b) Line M passes through points B (2, 5, 7) and C(−1, 3, 8). Find its vector equation.

Solution

a) We put in the position vector of the point we know on the line as **a** in the equation, and the vector telling us the direction as **d**.

NB: We must change coordinates into position vectors first!

So we get: $\mathbf{r} = 5\mathbf{i} + 3\mathbf{j} - 2\mathbf{k} + \lambda(4\mathbf{i} - 2\mathbf{j} - 3\mathbf{k})$

b) **Step 1** Work out the vector going between the two points on the line. This is **d**

We want $\overrightarrow{BC}$ (or $\overrightarrow{CB}$)
$\overrightarrow{BC} = -3\mathbf{i} - 2\mathbf{j} + \mathbf{k}$

Step 2 Use the position vector of one of the points as **a**, and put everything into the equation

We'll use B
$\mathbf{r} = 2\mathbf{i} + 5\mathbf{j} + 7\mathbf{k} + \mu(-3\mathbf{i} - 2\mathbf{j} + \mathbf{k})$

When we are looking at more than one line in a question, we use different letters for the parameter – we can't call it λ in more than one

Finding Whether a Point is on a Line

Example 3

The line L has equation $\mathbf{r} = 2\mathbf{i} + 3\mathbf{j} - \mathbf{k} + \lambda(-2\mathbf{i} + \mathbf{j} + 2\mathbf{k})$

Point P has position vector $6\mathbf{i} + \mathbf{j} - 5\mathbf{k}$. Point Q has position vector $5\mathbf{j} + \mathbf{k}$.

Determine whether points P and Q are on the line L.

Step 1 Replace **r** in the line equation with the position vector of point P

$6\mathbf{i} + \mathbf{j} - 5\mathbf{k} = 2\mathbf{i} + 3\mathbf{j} - \mathbf{k} + \lambda(-2\mathbf{i} + \mathbf{j} + 2\mathbf{k})$

Step 2 Equate **i**, **j**, and **k** components separately

i: $6 = 2 - 2\lambda$
j: $1 = 3 + \lambda$
k: $-5 = -1 + 2\lambda$

Step 3 Find the value of λ that each equation gives. If they are all the same, the point is on the line

i eqn: $\lambda = -2$
j eqn: $\lambda = -2$
k eqn: $\lambda = -2$ So point is on the line

Now for Q:

1 $5\mathbf{j} + \mathbf{k} = 2\mathbf{i} + 3\mathbf{j} - \mathbf{k} + \lambda(-2\mathbf{i} + \mathbf{j} + 2\mathbf{k})$
2 $0 = 2 - 2\lambda$; $5 = 3 + \lambda$; $1 = -1 + 2\lambda$
3 $\lambda = 1$; $\lambda = 2$; $\lambda = 1$ So point is not on the line

Deciding Whether Lines Intersect

Two lines in 3-dimensional space can intersect, be parallel or be skew.

The lines are parallel if their direction vectors are multiples of one another

(e.g. $2\mathbf{i} + \mathbf{j} - 3\mathbf{k}$ and $-4\mathbf{i} - 2\mathbf{j} + 6\mathbf{k}$, where the second vector is $-2 \times$ the first vector)

You should check whether lines are parallel before you start!

Example 4 shows how to decide whether lines intersect or are skew.

Example 4

Lines L and M have vector equations $\mathbf{r} = 2\mathbf{i} - \mathbf{j} + 6\mathbf{k} + \lambda(4\mathbf{i} + \mathbf{j} + 2\mathbf{k})$ and

$\mathbf{r} = -2\mathbf{i} - 4\mathbf{j} + 12\mathbf{k} + \mu(2\mathbf{i} + \mathbf{j} - \mathbf{k})$ respectively. Determine whether lines L and M are parallel, skew or intersecting. If they are intersecting, find the coordinates of their point of intersection.

Solution

Step 1 Check whether lines are parallel

Lines are not parallel, since $4\mathbf{i} + \mathbf{j} + 2\mathbf{k}$ and $2\mathbf{i} + \mathbf{j} - \mathbf{k}$ are not multiples of each other

Step 2 Put the two equations equal, and equate **i**, **j** and **k** components separately

$2\mathbf{i} - \mathbf{j} + 6\mathbf{k} + \lambda(4\mathbf{i} + \mathbf{j} + 2\mathbf{k})$
$= -2\mathbf{i} - 4\mathbf{j} + 12\mathbf{k} + \mu(2\mathbf{i} + \mathbf{j} - \mathbf{k})$
i: $2 + 4\lambda = -2 + 2\mu$ **(1)**
j: $-1 + \lambda = -4 + \mu$ **(2)**
k: $6 + 2\lambda = 12 - \mu$ **(3)**

Step 3 Use two equations to work out λ and μ (any two!)

Use first two equations:
(1) $\Rightarrow \mu = 2 + 2\lambda$
Sub. into **(2)** $\Rightarrow -1 + \lambda = -4 + 2 + 2\lambda$
$\Rightarrow \lambda = 1, \mu = 4$

Step 4 Substitute these values into third equation to see if it works. If it does, the lines intersect (and if not , they are skew)

(3) $6 + 2\lambda = 12 - \mu$. Put $\lambda = 1, \mu = 4$:
$6 + 2 \times 1 = 12 - 4$
$8 = 8$✔
So the lines intersect

Step 5 Find the point of intersection by putting either λ or μ back into the original line equation

Using L equation (because it's easier!):
$\mathbf{r} = 2\mathbf{i} - \mathbf{j} + 6\mathbf{k} + \lambda(4\mathbf{i} + \mathbf{j} + 2\mathbf{k})$
$\mathbf{r} = 2\mathbf{i} - \mathbf{j} + 6\mathbf{k} + 1(4\mathbf{i} + \mathbf{j} + 2\mathbf{k})$
$= 6\mathbf{i} + 8\mathbf{k}$
So its coordinates are (6, 0, 8)

Finding the Angle Between two Lines

To find the angle between two lines, you just find the angle between their direction vectors. However, if this is more than 90°, you must take it away from 180° to find the actual acute angle.

You ONLY do this with lines – NOT when finding the angle between vectors

Example 5

Find the angle between lines L and M in Example 4.

Solution

Direction vectors are $4\mathbf{i} + \mathbf{j} + 2\mathbf{k}$ and $2\mathbf{i} + \mathbf{j} - \mathbf{k}$

$(4\mathbf{i} + \mathbf{j} + 2\mathbf{k}).(2\mathbf{i} + \mathbf{j} - \mathbf{k}) = \sqrt{(4^2 + 1^2 + 2^2)}\sqrt{(2^2 + 1^2 + (-1)^2)}\cos\theta$

$8 + 1 - 2 = \sqrt{21}\sqrt{6}\cos\theta \Rightarrow \theta = 51.4°$

Applications Using Lines

Questions often require you to find a point on a line with certain properties. The key thing to remember is that you use the equation of the line as the position vector of the required point, then use whatever information you are given to calculate λ, and hence the actual point you need.

Example 6

The line L has equation $\mathbf{r} = 2\mathbf{i} + 3\mathbf{j} - 4\mathbf{k} + \lambda(2\mathbf{i} - \mathbf{j} - 3\mathbf{k})$

The point A has position vector $4\mathbf{i} + \mathbf{j} - 2\mathbf{k}$

a) Point P is on the line, and is such that angle AOP is 90°. Find P

b) Point Q is the closest point to A on the line. Find the distance AQ.

Solution

a) Whenever an angle is mentioned in vectors, we think 'dot product'.
For angle AOP, we need to look at the dot product of $\overrightarrow{OP}$ and $\overrightarrow{OA}$ (or $\overrightarrow{PO}$ and $\overrightarrow{AO}$)

Always do this in this type of question

Since P is on the line, we put $\overrightarrow{OP} = 2\mathbf{i} + 3\mathbf{j} - 4\mathbf{k} + \lambda(2\mathbf{i} - \mathbf{j} - 3\mathbf{k})$
$\overrightarrow{OA}$ = position vector of A = $4\mathbf{i} + \mathbf{j} - 2\mathbf{k}$

We need to simplify these by putting all the **i**'s together, all the **j**'s together, etc.
So $\overrightarrow{OP} = (2 + 2\lambda)\mathbf{i} + (3 - \lambda)\mathbf{j} + (-4 - 3\lambda)\mathbf{k}$

Angle AOP = $90° \Rightarrow \overrightarrow{OP}.\overrightarrow{OA} = 0$
So $4(2 + 2\lambda) + 1(3 - \lambda) - 2(-4 - 3\lambda) = 0$
$8 + 8\lambda + 3 - \lambda + 8 + 6\lambda = 0$

$19 + 13\lambda = 0 \Rightarrow \lambda = -\frac{19}{13}$

NOT the same value of λ as in part a)

Now we must substitute this value of λ back to find P: $\overrightarrow{OP} = \frac{1}{13}(-12\mathbf{i} + 58\mathbf{j} + 5\mathbf{k})$

b) We write $\overrightarrow{OQ} = 2\mathbf{i} + 3\mathbf{j} - 4\mathbf{k} + \lambda(2\mathbf{i} - \mathbf{j} - 3\mathbf{k})$

As we're interested in the distance between A and Q, we find $\overrightarrow{AQ}$
$\overrightarrow{AQ} = \overrightarrow{OQ} - \overrightarrow{OA} = 2\mathbf{i} + 3\mathbf{j} - 4\mathbf{k} + \lambda(2\mathbf{i} - \mathbf{j} - 3\mathbf{k}) - (4\mathbf{i} + \mathbf{j} - 2\mathbf{k})$
$= (-2 + 2\lambda)\mathbf{i} + (2 - \lambda)\mathbf{j} + (-2 - 3\lambda)\mathbf{k}$

The distance between A and Q is
$|\overrightarrow{AQ}| = \sqrt{[(-2 + 2\lambda)^2 + (2 - \lambda)^2 + (-2 - 3\lambda)^2]} = \sqrt{[14\lambda^2 + 12]}$
This is a minimum when $\lambda = 0$, since $\lambda^2 \geq 0$. So minimum distance is $\sqrt{12}$

If it had been something like $14\lambda^2 + 2\lambda + 12$, we'd have had to use completing the square or differentiation to find the value of λ giving a minimum

NB. We are not actually asked for Q – but if we were, we'd substitute in $\lambda = 0$

Vectors

30 minutes

Use your knowledge

1 The vector $\mathbf{v} = a\mathbf{i} + b\mathbf{j} + c\mathbf{k}$.
$\mathbf{v}$ is perpendicular to the vectors $\mathbf{i} - 2\mathbf{k}$ and $2\mathbf{i} + \mathbf{j} - \mathbf{k}$

a) Find a and b in terms of c.

Use the dot product

b) Given that $|\mathbf{v}| = \sqrt{56}$, and that c is positive, find c.

Write v in terms of c
Find |v| in terms of c, and equate to $\sqrt{56}$

2 The line L has equation $\mathbf{r} = (2 + \lambda)\mathbf{i} + (4 - 2\lambda)\mathbf{j} + (3\lambda - 1)\mathbf{k}$
The point P has position vector $-5\mathbf{i} + \mathbf{k}$.

a) Find a vector equation of the line OP.

You know $\overrightarrow{OP}$ already, and O is a point on the line

b) Find the angle between OP and L.

Get the equation of L into standard form first, by separating out the parts with λ in.
Remember you need to find the acute angle

3 The line M has equation $\mathbf{r} = 5\mathbf{i} + 2\mathbf{j} + \mu(\mathbf{i} - \mathbf{j} - \mathbf{k})$

a) i) The point Q lies on the line M, and is such that OQ is perpendicular to M.
Find the position vector of Q.

You need OQ to be perpendicular to the direction vector of the line

ii) Deduce the shortest distance between M and the origin.

"Deduce" means use a)i). Draw a sketch of O, M and Q to help you

b) The point R is the reflection of O in the line M. Find the position vector of R.

Draw a sketch

Test your knowledge

(In this chapter the acceleration (assumed constant) due to gravity; $g = 9.8\,\text{ms}^{-2}$.)

1 i) A particle P is projected horizontally with speed $23\,\text{ms}^{-1}$ from a point A, 44.1 m above a horizontal ground. The particle P moves freely under gravity and hits the ground at the point B. Calculate:

a) the time taken for the particle to travel from A to B
b) the horizontal distance between the points A and B
c) the speed and the angle with which the particle hits the ground.

ii) A different particle Q is projected horizontally from a point A, 40 metres above a horizontal ground. The particle Q moves freely under gravity and hits the ground at the point B, where the horizontal displacement from A to B is 60 m. Calculate to 2 significant figures:

a) the initial speed of the particle Q
b) the time taken for the particle Q to travel from A to B.

2 A stone is thrown from a point A with speed of $26\,\text{ms}^{-1}$, at an angle of elevation $\theta°$, where $\tan\theta = \frac{5}{12}$. The point A is 27.5 m above the horizontal ground level. The stone moves freely under gravity, and hits the ground level for the first time at the point B. Assuming that the air resistances are negligible, calculate:

a) the time taken for the ball to travel from A to B
b) the horizontal distance between the points A and B
c) the speed of the stone, 3 seconds after it has been thrown.

3 A particle P is projected from a point A on the ground, with speed $u\,\text{ms}^{-1}$, at an angle of elevation of $\theta°$, and moves freely under gravity. Show that the horizontal distance, R (i.e. the range), from A that the particle moves before it hits the ground is given by the expression: $R = \dfrac{2u^2 \sin\theta \cos\theta}{g}$

Answers

1 i) a) 3s **b)** 69m **c)** $37.3\,\text{ms}^{-1}$, 52.0° **ii) a)** $21\,\text{ms}^{-1}$ **b)** 2.86s

2 a) 3.60s **b)** 86.4m **c)** $30.9\,\text{ms}^{-1}$

3) AB $\rightarrow$: $s = R$, $u = u\cos\theta$, $t = t \Rightarrow R = (u\cos\theta) \times t$ **(1)**

AB $\uparrow$: $s = 0$, $u = u\sin\theta$, $a = -g$, $t = t \Rightarrow 0 = t(u\sin\theta - \frac{1}{2}gt) \Rightarrow$ At A, $t = 0$,
but:
At B: $u\sin\theta - \frac{1}{2}gt = 0 \Rightarrow t = \frac{2u\sin\theta}{g}$ **(2)**. Hence **(2)** into **(1)** gives answer.

✔ If you got them all right, skip to page 77

Projectiles

Improve your knowledge

A projectile is a particle that is projected with an initial velocity and then moves freely under gravity.

When your are answering projectile questions, you must split up the motion of a projectile into two separate components: a horizontal velocity component and a vertical velocity component.

Since gravity acts downwards, towards the centre of the Earth, the vertical component of velocity is subject to constant acceleration. This means that the projectile's vertical component of velocity is always changing. Hence for the vertical component of velocity, we need to use any of the equations of motion with constant acceleration.

i.e.: From *AS Maths*: $v = u + at$ $\quad s = ut + \frac{1}{2}at^2$ $\quad v^2 = u^2 + 2as$

There is no force acting on a particle horizontally. This means that the particle's horizontal component of velocity is always the same. When resolving horizontally we know the particles acceleration is $0\,ms^{-2}$, which means the equation of motion $s - ut + \frac{1}{2}at^2$, becomes: $s = ut$

Key points from AS in a Week

Equations of Motion pages 46–48

Horizontal Projection

Example 1

A particle P is projected horizontally with speed $16\,ms^{-1}$ from a point A, 78.4 m above a horizontal ground. The particle P moves freely under gravity and hits the ground at the point B. Calculate:

a) the time taken for the particle to travel from A to B
b) the horizontal displacement between the points A and B
c) the speed of the particle, 1.5 seconds after leaving the point A.

Solution

Draw a diagram to help you.

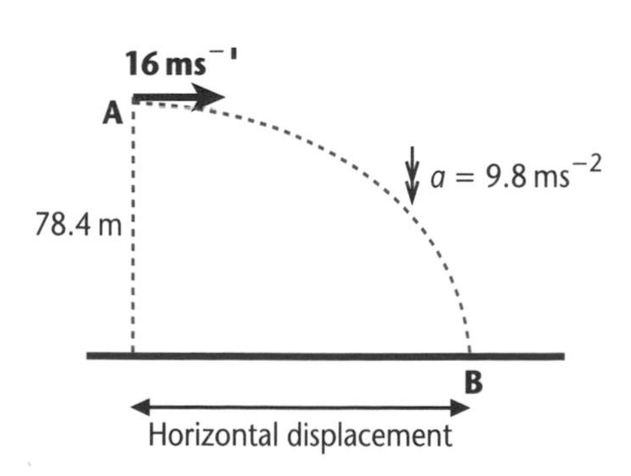

a) Resolve vertically ↓ (+ve) A to B

$u = 0$ (vertical velocity is zero)
$s = 78.4$ (vertically)
$a = 9.8$
$t = ??$
$v = ??$

Choose formula: $s = ut + \frac{1}{2}at^2 \Rightarrow 78.4 = (0 \times t) + \frac{1}{2}(9.8)t^2$

Solve: $78.4 = 4.9t^2 \Rightarrow t = \sqrt{\frac{78.4}{4.9}} = 4\,\text{s}$

NB: The time taken for the particle to travel vertically from A to B is the same as the time taken for the particle to move horizontally from A to B.

b) Resolve horizontally (→) A to B

$s = ??$

$u = 16$

$t = 4$ (from part a)

Use formula: $s = ut \Rightarrow s = 16 \times 4 = 64\,\text{m}$

c) Assuming there is no air resistance, the horizontal component of the particle's velocity during its motion will remain the same. After 1.5 seconds, $v_H = 16\,\text{ms}^{-1}$

Vertical velocity changes, due to gravity ⇒ Resolve and use equations of motion:

Resolve vertically (↓) from A:

$u = 0$

$s = ??$

$a = 9.8$

$t = 1.5$

$v = ??$

Formula, need v, so: $v = u + at$

Apply: $v_v = (0) + (9.8 \times 1.5) = 14.7\,\text{ms}^{-1}$

Find the speed using Pythagoras' Theorem:

speed $= \sqrt{(16)^2 + (14.7)^2}$

$= 21.728... = 21.7\,\text{ms}^{-1}$ (3sf)

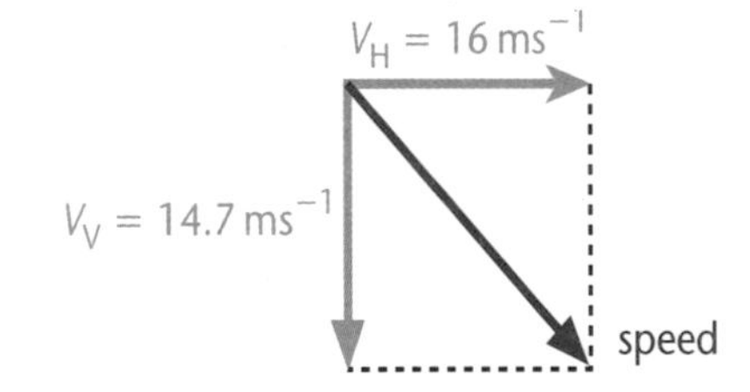

Example 2

A particle P is projected horizontally from a point A, 30 metres above a horizontal ground. The particle P moves freely under gravity and hits the ground at the point B, where the horizontal displacement from A to B is 50 metres. Calculate to 2sf:

a) the initial speed of the particle P

b) the time taken for the particle to travel from A to B.

Solution

Draw a diagram to help you!!

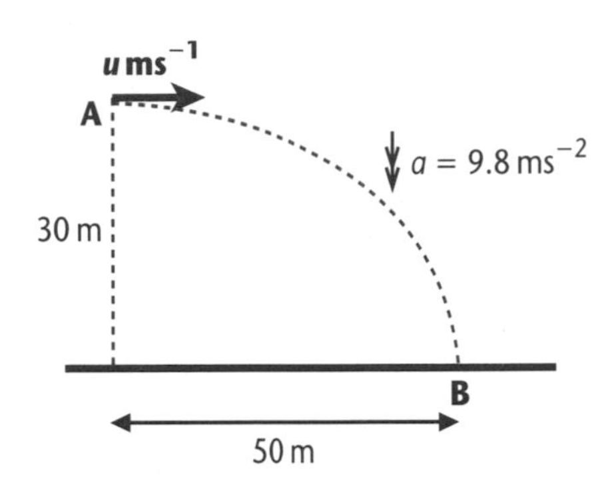

a) Resolve vertically (↓) A to B

$u = 0$ (zero vertical velocity)
$s = 30$
$a = 9.8$
$t = ??$
$v = ??$

Resolve horizontally (→) A to B

$s = 50$
$u = u$
$t = ??$

Formula: $s = ut \Rightarrow 50 = ut$
Rearrange: $t = \frac{50}{u}$

The time, $t = \frac{50}{u}$, taken to travel horizontally from A to B is the same as the time taken to travel vertically from A to B. So $t = \frac{50}{u}$ is the common link.

Vertically from A to B with $t = \frac{50}{u} \Rightarrow$ Formula: $s = ut + \frac{1}{2}at^2$

Apply: $30 = \left(0 \times \frac{50}{u}\right) + \left(\frac{1}{2} \times 9.8 \times \left(\frac{50}{u}\right)^2\right) \Rightarrow 30 = \frac{12250}{u^2}$

Rearrange: $u = \sqrt{\frac{12250}{30}} = 20.207... = 20\text{ms}^{-1}$ (2sf)

b) Using: $t = \frac{50}{u} \Rightarrow t = \frac{50}{20.207...} \Rightarrow 2.4744... = 2.5\text{s}$ (2sf)

2 Inclined Projection

If a particle is projected, with velocity $u\text{ms}^{-1}$, at an angle of $\theta°$ to the horizontal, then its initial velocity can be split into horizontal and vertical components, with:

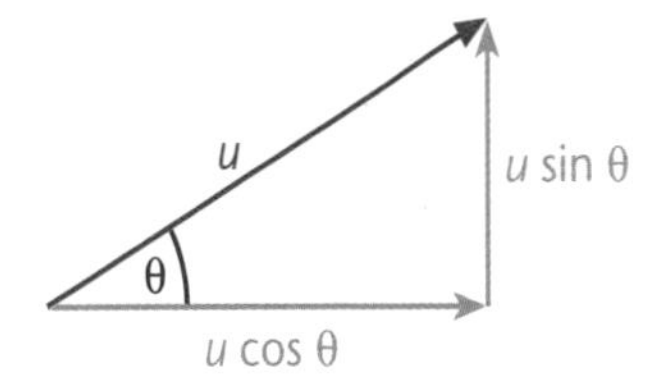

$u_H = u\cos\theta$ and $u_V = u\sin\theta$

Example 3

A golfer hits a golf ball, on a horizontal shooting range, with a velocity of 50ms^{-1} at an angle of 25° above the horizontal. Assuming that the golf ball can be modelled as a particle moving freely under gravity, calculate the:

a) greatest height reached by the golf ball
b) time taken by the golf ball to reach this greatest height

c) time taken by the golf ball to first hit the ground
d) horizontal distance travelled by the golf ball when it first hits the ground.

State two physical factors that would need to be considered to make the modelling of this problem more realistic.

Solution

a) Draw a diagram to help you!!

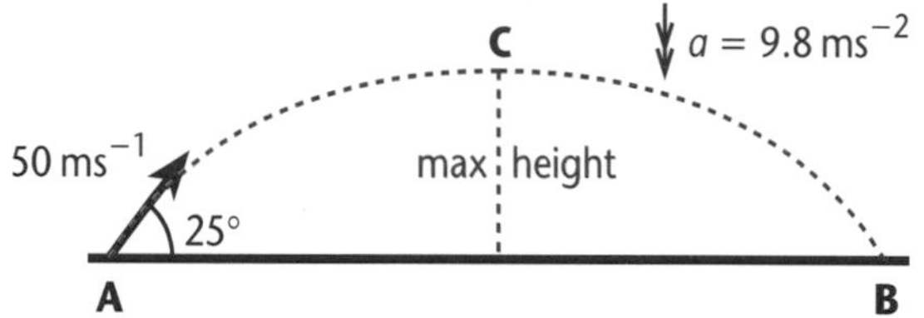

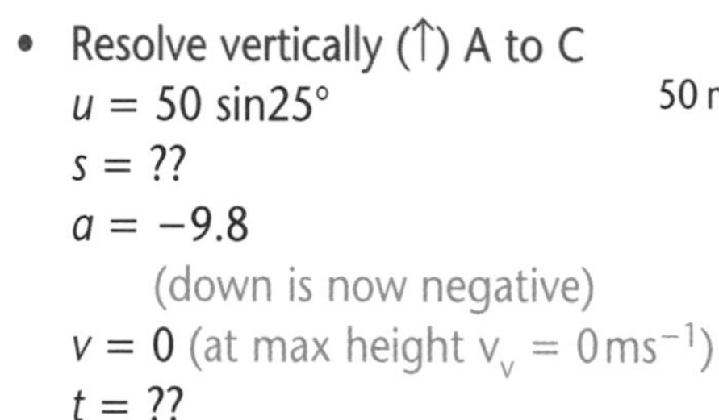

- Resolve vertically (↑) A to C
  $u = 50\sin 25°$
  $s = ??$
  $a = -9.8$
  (down is now negative)
  $v = 0$ (at max height $v_v = 0\,ms^{-1}$)
  $t = ??$
- Need s: Formula $\Rightarrow v^2 = u^2 + 2as \Rightarrow 0^2 = (50\sin 25)^2 + 2(-9.8)s$
- Rearrange: $19.6s = (50\sin 25)^2 \Rightarrow s = \dfrac{(50\sin 25)^2}{19.6} = 22.781 = 22.8\,m$ (3sf)

b) Need t: Formula $\Rightarrow v = u + at \Rightarrow 0 = (50\sin 25) + (-9.8)t$

Rearrange: $9.8t = 50\sin 25 \Rightarrow t = \dfrac{50\sin 25}{9.8} = 2.1562... = 2.16\,s$ (3sf)

c) By looking at the diagram, the path of the ball as it moves from A to B is symmetrical about the maximum height C. This means that the time taken for the ball to hit the ground is twice the time taken for the ball to reach its maximum height. $\therefore$ Time taken $= 2 \times 2.1562... = 4.3124... = 4.31\,s$ (3sf)

d) Resolve horizontally (→) A to C

$s = ??$
$u = 50\cos 25°$
$t = 4.3124...$

Formula: $s = ut$
$s = (50\cos 25°) \times 4.3124...$
$s = 195.42... = 195\,m$ (3sf)

e) Examples include: air resistance, variation in g, wind, spin of ball, tee off ground.

Example 4

A particle P is projected from a point A with speed of $10\,ms^{-1}$, at an angle of elevation $\theta°$, where $\tan\theta = \frac{3}{4}$. The point A is 54.4 m above the horizontal ground level. The particle P moves freely under gravity, and hits the ground level for the first time at the point B. Assuming, that the air resistances are negligible, calculate:

a) the time taken for the ball to travel from A to B
b) the horizontal distance between the points A and B.

Solution

Since $\tan\theta = \dfrac{opp}{adj} = \dfrac{3}{4}$,

construct a 3,4,5 right-angled triangle

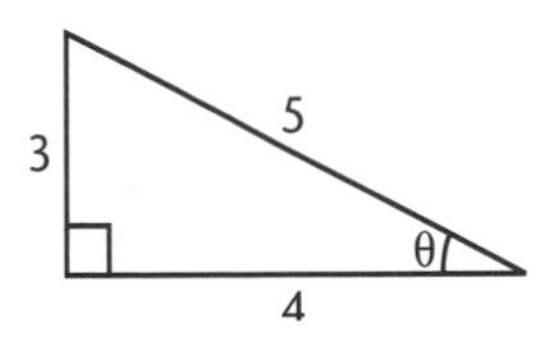

Hence: $\sin\theta = \dfrac{opp}{hyp} = \dfrac{3}{5}$ and $\cos\theta = \dfrac{adj}{hyp} = \dfrac{4}{5}$

a) Draw a diagram to help you.
Taking A as the origin:
Resolve vertically (↑) A to B:
$u = 10\sin\theta = 10 \times \frac{3}{5} = 6$
$s = -54.4$ (B is 54.4 m below A)
$a = -9.8$ (since up is positive)
$t = ??$
$v = ??$

Need t: Formula $\Rightarrow s = ut + \frac{1}{2}at^2 \Rightarrow -54.4 = 6t + \frac{1}{2}(-9.8)t^2$

Form to quadratic: $-54.4 = 6t - 4.9t^2 \Rightarrow 4.9t^2 - 6t - 54.4 = 0$

Quadratic formula: $$t = \frac{-(-6) \pm \sqrt{36 - (4 \times 4.9 \times -54.4)}}{2 \times 4.9}$$

$$= \frac{6 \pm \sqrt{1102.24}}{9.8} = \frac{6 \pm 33.2}{9.8}$$

Hence: $t = -2.78$ s (reject) and $t = 4$ s (accept!)

b) Resolve horizontally (→) A to B
$s = ??$
$u = 10\cos\theta = 10 \times \frac{4}{5} = 8$
$t = 4$ (from part a))

Formula: $s = ut$
$s = 8 \times 4$
$s = 32$ m

3 General Formulae

Some examination questions may ask you to derive general formulas for the motion of a projectile.

Example 5

A particle P is projected from a point A on the ground, with speed $u\,\text{ms}^{-1}$ at an angle of elevation $\theta°$. Show that the maximum height reached by the particle is given by the expression $\frac{u^2\sin^2\theta}{2g}$, where g is the acceleration due to gravity.

Solution

- Draw a diagram to help you.
- Let the maximum height reached be denoted as H.

Projectiles

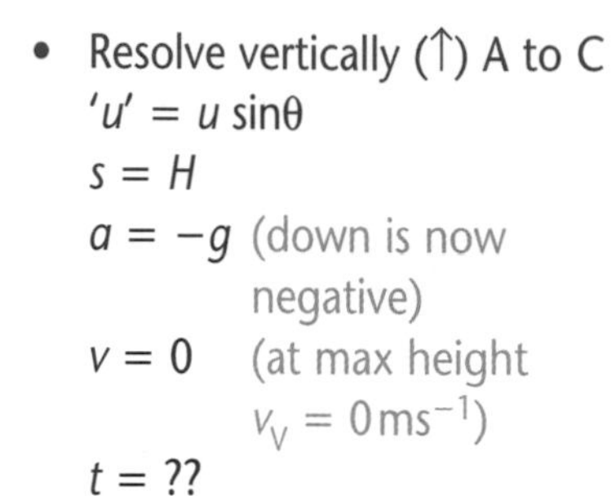

- Resolve vertically (↑) A to C
  $'u' = u \sin\theta$
  $s = H$
  $a = -g$ (down is now negative)
  $v = 0$ (at max height $v_V = 0\,\text{ms}^{-1}$)
  $t = ??$

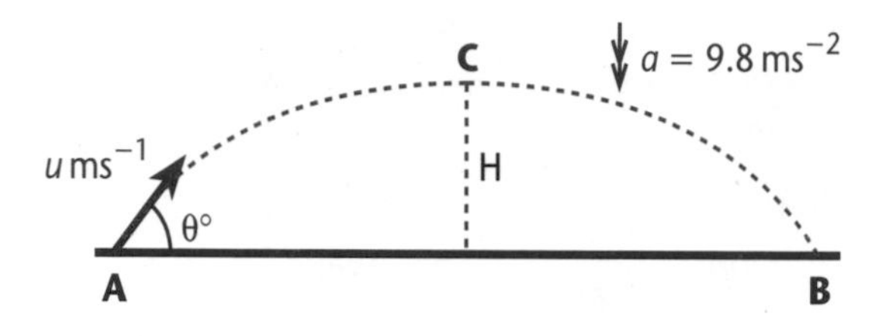

- Need s: Formula $\Rightarrow v^2 = u^2 + 2as \Rightarrow 0^2 = (u \sin\theta)^2 + (2 \times -g \times H)$
- Rearrange: $0 = u^2 \sin^2\theta - 2gH \Rightarrow 2gH = u^2 \sin^2\theta$
- Hence: $H = \frac{u^2 \sin^2\theta}{2g}$ (as required)

35 minutes

Use your knowledge

1 A small pebble is thrown from the top of a vertical cliff, which is 80 metres above sea level. The pebble is thrown with an initial velocity of $12\,ms^{-1}$, at an angle of depression of 15°. Assuming that the stone is modelled as a particle moving freely under gravity, find:

a) the time taken for the pebble to hit the sea

Form a quadratic in t

b) the horizontal distance from the base of the cliff to where the pebble hits the sea.

Use $s = ut$, horizontally

2 A particle is projected from a point A on horizontal ground such that it just clears a vertical wall of height 3.97 m, at the top of the particle's trajectory. Given that the wall is 21.2 m from the point A, as shown in the diagram:

Top of particle's trajectory, ⇒ vertical velocity = $0\,ms^{-1}$

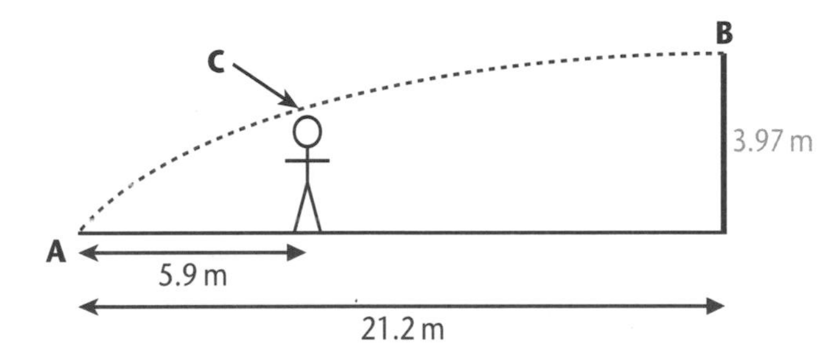

a) Calculate, to 3 significant figures, the time taken for the particle to just clear the top of the wall.

Find vertical u, then find t

A man of height 1.8 m stands between A and the wall, at a distance of 5.9 m horizontally from A.

b) Verify that the particle will pass over the man.

Find horizontal u. Use $s = ut$ to find time to AC. Then resolve AC(↑)

3 A particle is projected from a point A, on horizontal ground with velocity, $\mathbf{v} = (26\mathbf{i} + 15\mathbf{j})\,ms^{-1}$. After the particle reaches its maximum height, it hits a vertical wall at the point B, which is 6.9 metres above the ground. Calculate, to 3 significant figures:

$\therefore u_H = 26\,ms^{-1}$, $u_V = 15\,ms^{-1}$

a) the maximum height reached by the particle

Max height ⇒ $V_V = 0\,ms^{-1}$

b) the time taken for the particle to hit the wall

AB (↑), then quadratic in t

c) the speed at which the particle hits the wall.

Use Pythagoras

Work, Energy and Power

Test your knowledge

Take g = 9.8 ms^{-2} throughout this chapter.

A ball of mass 250 g is thrown vertically upwards with speed 30 ms^{-1}.

a) Use energy considerations to find the maximum height to which it rises.
b) Find its speed at the instant when it is 10 m above the ground.

a) A particle of mass 0.5 kg is acted on by a force so that its speed increases from 2 ms^{-1} to 5 ms^{-1}. Find the work done by the force.
b) A boy of mass 20 kg on a scooter of mass 5 kg is standing on a road inclined at 10° to the horizontal. His father pushes him for 30 seconds; during that period he moves 50 m up the road and develops a speed of 8 ms^{-1}. Find the average force exerted by the boy's father.

a) A disc of mass 10 grams is given a push so that it slides along a horizontal surface. It comes to rest in 80 cm. The coefficient of friction between the disc and the surface is 0.8. Use energy considerations to find the initial speed of the disc.
b) A particle of mass 1 kg is at rest on a plane inclined at 20° to the horizontal. It is projected up the plane with speed 30 ms^{-1}. The average resistive force acting on the particle is 3 N. Find the distance up the plane that the particle moves.

a) A car of mass 1000 kg has a maximum speed of 17 ms^{-1} on a level road. The frictional resistance to motion of the car is 1500 N. Find the power of the car's engine.
b) The car is now accelerating up a hill inclined at $\sin^{-1}(0.01)$ to the horizontal. The frictional resistance to motion and the power of the engine are unchanged. Find the rate at which the car is accelerating at the instant when its speed is 10 ms^{-1}.

Answers

1 a) 45.9 m **b)** 26.5 ms^{-1}
2 a) 5.25 J **b)** 58.5 N
3 a) 3.54 ms^{-1} **b)** 70.8 m
4 a) 25.5 kW **b)** 0.952 ms^{-2}

If you got them all right, skip to page 83

Work, Energy and Power

Improve your knowledge

Conservation of Energy

If no external forces are acting, total energy (kinetic + potential energy) is conserved 'No external forces' means that you can't have friction or air resistance – a body's weight does not count as an external force.

The best way to use this is to say 'total energy before = total energy after'.

Example 1

A ball of mass 0.05 kg is thrown vertically downwards from a cliff with speed $20\,\text{ms}^{-1}$. As it hits the sea at the bottom of the cliff, it has speed $50\,\text{ms}^{-1}$.

Find the height of the cliff.

Solution

Step 1 Decide where to take potential energy to be zero

p.e. = 0 at top of cliff

You can choose anywhere you like!

Step 2 Work out total initial energy

Initial k.e. $= \frac{1}{2}mv^2 = \frac{1}{2} \times 0.05 \times 20^2$
Initial p.e. $= 0$
So total initial energy = 10 J

Step 3 Work out total final energy

Final k.e. $= \frac{1}{2} \times 0.05 \times 50^2$
Final p.e. $= mgh = 0.05 \times 9.8 \times -h$
So total final energy $= 62.5 - 0.49h$

Potential energy is negative as it is below the top of the cliff

Step 4 Equate and solve

$10 = 62.5 - 0.49h$
$h = 107\,\text{m}$

Work Done by a Force

A force does work when it moves an object. The formula is:

Work done by a force = Force × Distance moved **in the direction of the force**

or: = Distance moved × Component of force in direction moved

You also need to know:

Work done by a force = Increase in total energy = Final energy − Initial energy

You must be careful to distinguish between work done by a force – when the force is 'helping' something to move, and so increasing the total energy – and work done against a force, when the force is trying to stop something moving (like friction) and so decreasing the total energy.

Questions usually rely on you combining the above two equations.

Example 2

A car of mass 800 kg is at rest on a road inclined at 2° to the horizontal. Five seconds later, it is 100 m further up the road, travelling at $10\,ms^{-1}$. Assuming there are no frictional resistances to motion, find the average driving force exerted by the car's engine.

Solution

Step 1 Draw a diagram and mark on the zero level for potential energy

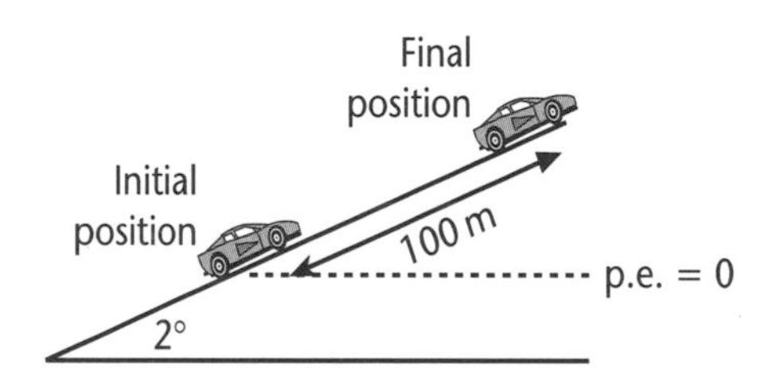

Step 2 Work out initial and final total energy

Total initial energy = 0

Final k.e. = $\frac{1}{2} \times 800 \times 10^2$

Final p.e. = $800 \times 9.8 \times 100\sin 2°$

Total final energy = 67 361 J

You must use vertical distance in potential energy formula

Step 3 Use the two formulae for work done and equate

Work done by force = Final – Initial
= 67 361 J

Work done = force × dist. moved in direction
= $D \times 100$

So $67\,361 = 100D \Rightarrow D = 674$ N

Driving force is acting directly up the hill, in the same direction as the distance moved

Work Done Against a Force

When a force is resisting motion (like friction), a body does work against it when it moves.

Work done against a force = Decrease in total energy = Initial – Final

Note this is the other way round!

Work done against a force = Force × Distance moved **opposite to direction of force**

or: = Distance × Compt of Force opposite direction moved

Questions work in very much the same way as those on work done by a force; the steps used are identical.

Example 3

A particle of mass 0.6 kg is held at rest at the top of a slope inclined at 1° to the horizontal. The frictional force acting on the particle due to the plane is 1 N. The particle is set moving down the plane with a speed of $2\,ms^{-1}$. Use energy considerations to determine how far the particle travels down the slope before coming to rest.

As friction is involved, it must be work against a force

Solution

1

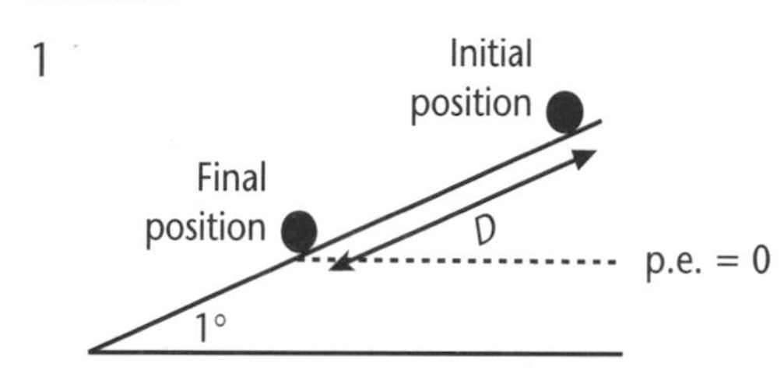

2 Initial potential energy $= 0.6 \times 9.8 \times D\sin 1°$
Initial kinetic energy $= \frac{1}{2} \times 0.6 \times 2^2$
Total initial energy $= 1.2 + 0.1026D$

Final potential energy $= 0$
Final kinetic energy $= 0$ (as it is at rest)
So total final energy $= 0$

3 Work done against friction = initial − final $= 1.2 + 0.1026D$
Work done against friction $= 1 \times D$
So $D = 1.2 + 0.1026D$
$0.8974D = 1.2 \Rightarrow D = 1.34\,m$

Power

Power is the rate of doing work. The key equation to remember is

Power = Driving force × Velocity

That's the driving force exerted by a vehicle – not the resultant force on the vehicle

Example 4

A car of mass 900 kg can travel at a maximum speed of $20\,ms^{-1}$ on a level road. The frictional resistance to motion of the car is 450 N.

a) Find the power of the car's engine.

The car is travelling up a road inclined at $\sin^{-1}(0.02)$ to the horizontal with its engine working at the same rate. At one instant, it is accelerating at $0.3\,ms^{-2}$.

'Working at the same rate' means exerting the same power

b) Find the speed of the car at this instant, given that the frictional resistance to motion is unchanged.

Work, Energy and Power

Solution

a) **Step 1** Draw a diagram, showing all forces

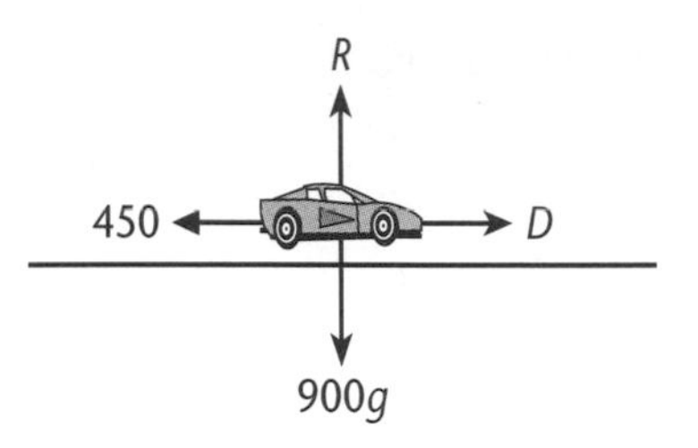

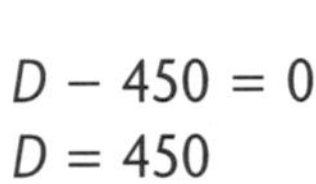

Step 2 Resolve in the direction of motion. Use $F = ma$

$D - 450 = 0$
$D = 450$

If the car is at maximum speed, there is no resultant force!

Step 3 Use the power equation

$P = Dv$
$= 450 \times 20 = 9000\text{W}$

b) **1**

R
D
450
900g
$\sin^{-1}(0.02)$

D is NOT the same as before! Call it D_1 if you think you might forget this. It's just the POWER that's the same

2 $D - 450 - 900g\,(0.02) = 900 \times 0.3$
$D = 896.4\text{N}$

3 $P = Dv \Rightarrow 9000 = 896.4v \Rightarrow v = 10.0\,\text{ms}^{-1}$

We now must resolve up the slope, as that's the direction the car is moving in

Use your knowledge

1 A car of mass 1200 kg is towing a trailer of mass 800 kg along a level road. The car and trailer are travelling at a constant speed of $20\,\text{ms}^{-1}$. The car's engine is working at a rate of 10 kW. The resistances to motion of the car and the trailer are proportional to their masses.

a) i) Find the resistance to motion of the car.

Consider the car and trailer separately. Don't forget to include the tension in the tow bar. Add together the two equations you get. Find D from the power. Find the total resistance, then work out what fraction of it belongs to the car

ii) Find the tension in the tow-bar.

Substitute into one of your equations from a) i)

The car and trailer start to travel down a slope inclined at 5° to the horizontal.

b) Find their acceleration at the instant their speed is $20\,\text{ms}^{-1}$. You may assume the car's engine exerts the same total power, and that frictional resistances to motion are unchanged.

Consider them separately again. They each have the same acceleration. Don't forget the component of the weight down the slope is helping, not hindering!

2 A particle of mass 60 grammes is held at rest on a plane inclined at 30° to the horizontal. The coefficient of friction between the particle and the plane is 0.1. The particle is projected up the plane with speed $4\,\text{ms}^{-1}$. Find the distance it has travelled when it first comes to rest.

As friction is involved, work is being done against a force. You need to use $F = \mu R$ to find the frictional force. Call the distance d, and get the potential energy and work done in terms of it

Test your knowledge

1 The particles A, B and C have masses m, $3m$ and $4m$ respectively, and are at coordinates $(2,-3)$, $(4,6)$ and $(1,3)$ respectively. Find the coordinates of their centre of mass.

2 Find the position of the centre of mass of the bodies shown.

a) AB, BC and AC are wires of uniform thickness and lengths $3a$, $4a$ and $5a$ respectively.

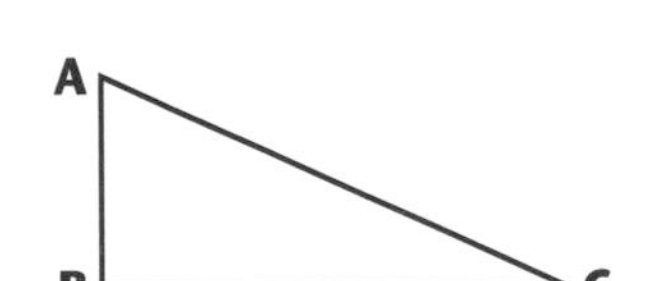

b) ABCDEF is a lamina of uniform thickness. $AB = BC = DE = 2a$, $CD = 3a$.

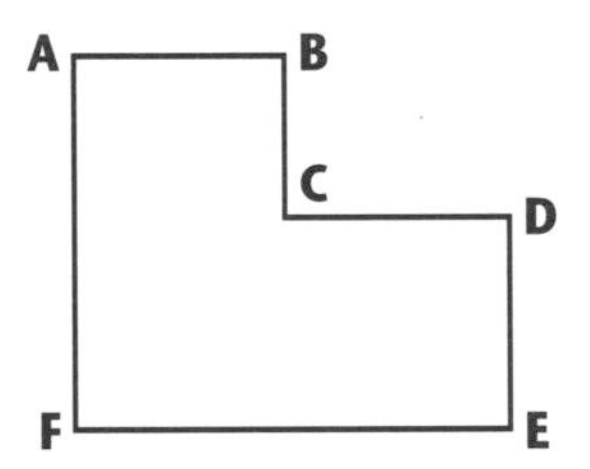

c) ABC is a triangular lamina of base $6a$ and height $9a$. A square of side $2a$ has been removed from the centre of its base, as shown.

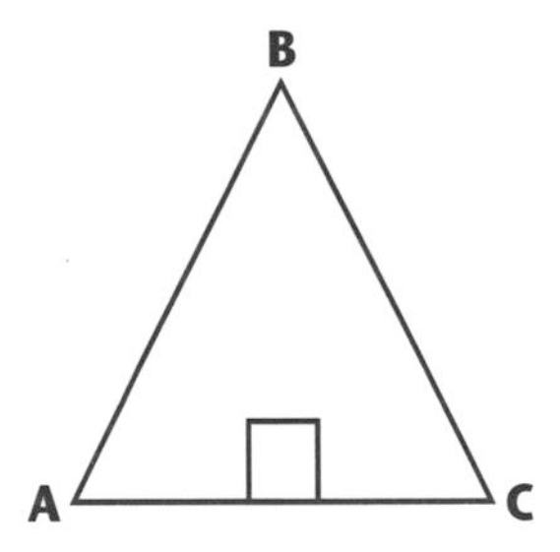

3 The lamina from 2 b) is suspended from point E, so that it hangs freely. Find the angle that EF makes with the vertical.

1 (2.25, 3.375)

2 a) Distance a from BC and $1.5a$ from AB **b)** $\frac{29a}{14}$ from AF, $\frac{11a}{7}$ from EF

c) $\frac{77a}{23}$ from the base, on the line of symmetry

3 28.2°

If you got them all right, skip to page 90

Improve your knowledge

The centre of mass of anything is the point at which we can consider its weight to act.

1 Centre of Mass of Particles

The centre of mass of a set of particles is found using:

Σ (mass × coordinates) = Total mass × coordinates of centre of mass

Example 1

a) Particles of mass m, $2m$ and $4m$ are at points A(1,5), B(3, −4) and C(7, −1) respectively. Find the centre of mass of the system.

b) A particle of mass $5m$ is added so that the centre of mass of the system is at (0,0). Find the position of the particle of mass $5m$

Solution

a) $m(1,5) + 2m(3, -4) + 4m(7, -1) = 7m(x, y)$
$(35m, -7m) = (7mx, 7my)$
$(5, -1) = (x, y)$

b) We can treat the original three masses as if they were one mass of $7m$ at (5, −1):
$7m(5, -1) + 5m(x, y) = 12m\,(0, 0)$
$(35m + 5mx, -7m + 5my) = (0, 0)$
So $35m + 5mx = 0 \Rightarrow x = -7$. $-7m + 5my = 0 \Rightarrow y = 1.4$

2 Centre of Mass of Wires & Laminas

Basic Facts

- The centre of mass of any uniform body is on all its axes of symmetry
- The centre of mass of a triangle is a perpendicular distance $\frac{1}{3}h$ from its base, where h is its perpendicular height

To find the centre of mass of any other body, we divide it up into pieces like rectangles or triangles, and use the formula:

- (mass of part × distance from a line) = Mass of whole × distance of centre of mass from line

You are usually asked to find the distance of the centre of mass from one or more lines; if you are just asked for its position, you choose the lines to measure from yourself and state which ones you used.

Centres of Mass

Always see if you can use symmetry to help you!

In addition, we'll use the facts that

- For a uniform wire, the mass of a piece is proportional to its length
- For a uniform lamina, the mass of a piece is proportional to its area

This lets us work with pieces of wire/parts of a lamina without knowing their actual masses.

Example 2

In the diagram below, AB = 4 cm, BC = 4 cm and CD = 1 cm

Exam hint: Always check carefully whether it's a lamina or a wire! The answers will be very different!

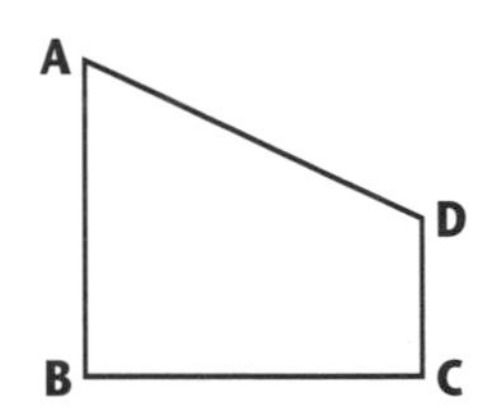

Find the distance of the centre of mass of the body from AB and BC if

a) the body is made of uniform wire
b) the body is a uniform lamina.

Solution

a) The best way to approach this sort of problem is to draw a table:
We need to know length of AD: by Pythagoras, AD = $\sqrt{(3^2 + 4^2)} = 5$ cm
We will use the length of each wire as its mass.

Body	Mass	Distance of centre of mass from	
		AB	BC
AB	4	0	2
BC	4	2	0
CD	1	4	0.5
AD	5	2	2.5
ABCD	14	a	b

Note the centre of mass of each wire is halfway along its length

To find the position of the centre of mass of AD, we use the fact that it must be half way along horizontally and half way along vertically

Now use the formula to work out the distance of the centre of mass from AB:

$4 \times 0 + 4 \times 2 + 1 \times 4 + 5 \times 2 = 14 \times a \Rightarrow a = \frac{11}{7}$

Now repeat for distances from BC:

$4 \times 2 + 4 \times 0 + 1 \times 0.5 + 5 \times 2.5 = 14 \times b \Rightarrow b = \frac{3}{2}$

b) In this case, we have to divide it up into a triangle and a rectangle, as shown:

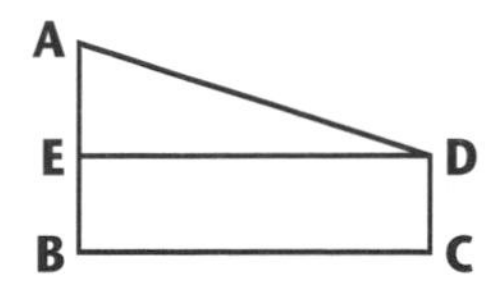

Body	Mass	Distance of centre of mass from	
		AB	BC
ADE	6	$\frac{4}{3}$	2
BCDE	4	2	0.5
ABCDE	10	a	b

Note the centre of mass of the triangle is one third of ED from AE, and one third of AE from ED

$$6 \times \tfrac{4}{3} + 4 \times 2 = 10 \times a \Rightarrow a = 1.6$$

$$6 \times 2 + 4 \times 0.5 = 10 \times b \Rightarrow b = 1.4$$

Sometimes the object for which we need to find the centre of mass is formed by a piece being cut out of a lamina. The principle is the same here – but the 'whole thing' is the lamina before the piece was cut out, and the unknowns are in a different place in the table.

Example 3

Find the distance of the centre of mass of the body shown from the centre of the larger circle. The two circles have radii 8 cm and 3 cm.

Solution

Body	Mass	Distance of centre of mass from centre of large circle
Small Circle	9π	5
Lamina	55π	x
Large circle	64π	0

Keep things in terms of π – it is tidier!

$$9\pi \times 5 + 55\pi \times x = 64\pi \times 0 \Rightarrow x = -\tfrac{9}{11}$$

Note the minus sign here – this is because it is in the opposite direction – to the left of the centre

Using Centre of Mass

There are two main uses of centre of mass:

- When an object is hung up by a corner, the centre of mass is vertically below the point of suspension.
- When an object is on an inclined plane, it will not tip over as long as the centre of mass is above a part of the object in contact with the plane.

Examples 4 and 5 illustrate how these work.

Example 4

The uniform lamina from Example 2 b) is suspended from point A so that it hangs freely. Find the angle that AB makes with the vertical.

Solution

Step 1 Draw a diagram with the centre of mass (call it G) directly under the point of suspension. (The diagram doesn't have to be accurate!) Mark the angle you have to find

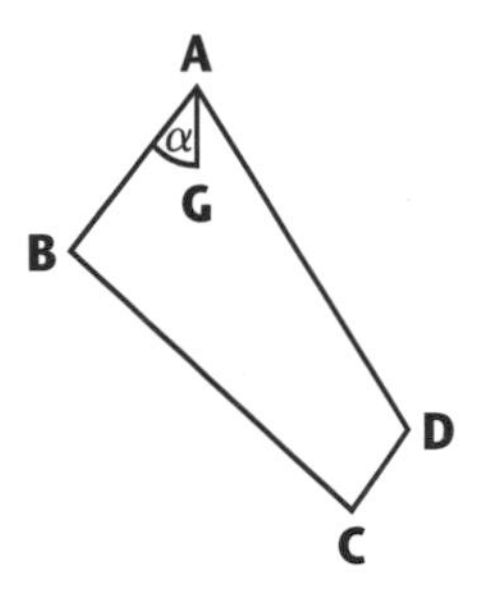

Step 2 Use the information you know about the position of the centre of mass to find $\tan\alpha$, and so α

We know G is 1.6 cm from AB and 1.4 cm from BC

So $\tan\alpha = \frac{1.6}{4 - 1.4} \Rightarrow \alpha = 31.6°$

Be careful! Don't just put the numbers you've got in without checking

Example 5

A uniform rectangular lamina of sides $3a$ and $4a$ is placed with a shorter side on a rough plane inclined at angle α to the horizontal. Find the maximum possible value of α, given that the lamina does not slip.

Solution

Step 1 Draw the diagram showing the lamina about to slip – so the centre of mass is directly above the lower corner

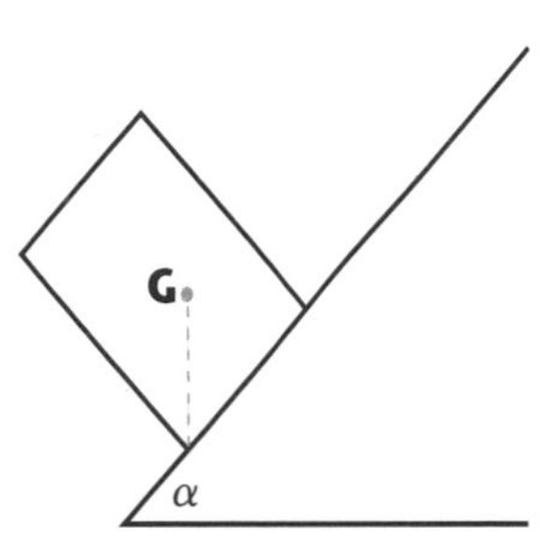

Step 2 Use what you know about the position of the centre of mass to work out α

G is in the centre of the lamina. We use angle facts to connect α and the measurements we know

So $\tan\alpha = \frac{1.5a}{2a} = 0.75$
$\Rightarrow \alpha = 38.9°$

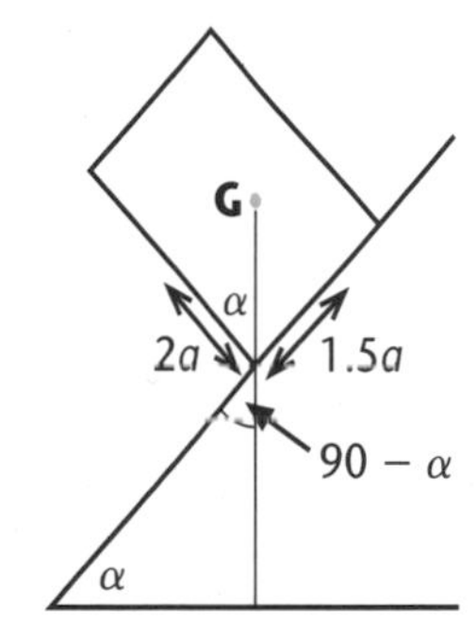

Centres of Mass

30 minutes

Use your knowledge

1 ABCD is a uniform rectangular lamina with AB = CD = 20 cm and BC = AD = 10 cm. Points E and F lie on AB and AD respectively and AE = AF = 5 cm. The lamina is folded along the line EF so that the triangle AEF is double thickness.

a) Find the position of the centre of mass of the resulting body.

Draw a diagram! Divide the lamina up into a triangle and two rectangles. Don't forget the triangle is double thickness, so double mass

b) The body is suspended from point B. Find the angle that BC makes with the vertical.

Draw a diagram! Centre of mass is directly below B

2 *In this question, you may use the fact that the centre of mass of a uniform semicircular lamina of radius a is at a distance of $\frac{4a}{3\pi}$ from its straight side.*

A semicircular lamina of radius $2R$ has a semicircular lamina of radius R removed from it, as shown below.

a) Find the position of the centre of mass of this lamina.

You need to make the "whole thing" the large semicircle. You need to look at distance from AB and from a line perpendicular to AB

b) The lamina is of mass M. A particle of mass $2M$ is attached at point B.

Find the position of the centre of mass of the combined object.

Treat the lamina as if all its mass is at its centre of mass, then combine the two in the normal way

Test your knowledge

Find the moment of each force about the point A.

a)

A 30 cm 4 N

b)

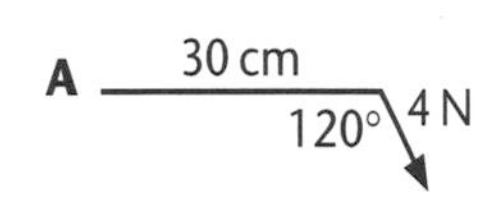

c)

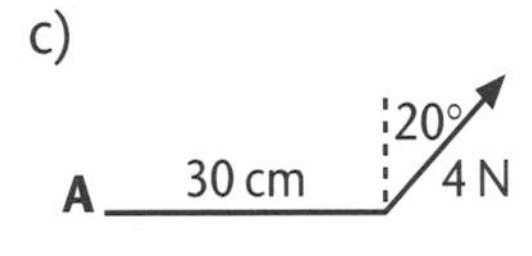

2

a) A smooth uniform plank of length 6m has mass 10kg. It rests on two supports, placed 1 m from each of its ends. A particle of mass 1 kg is placed 2m from one end of the plank. Find the magnitude of the reaction at each of the supports.

b) A non-uniform plank AB of length 3m has mass 5kg. A support is placed under the plank 1 m from end A. The plank will balance if a mass of 1 kg is placed on end A. Find the distance of the centre of mass of the plank from end A.

Answers

1 a) 1.2Nm clockwise **b)** 1.04Nm clockwise **c)** 1.13Nm anticlockwise

2 a) 5.25g, 5.75g **b)** 1.2m

If you got them all right, skip to page 94

Moments

Improve your knowledge

Moment of a force

The moment of a force is its turning effect. It can be clockwise or anticlockwise. Moments always have to be about a particular point – called the pivot. Deciding which point to take moments about will be covered later.

To find the moment of a force about a point, you use the formula:

Moment (in Nm) = distance between point of action of force and pivot (m) × component of force perpendicular to line along which distance is measured (N)

Pivot

Component in this direction required

Force

Distance

ALWAYS say whether it is clockwise or anticlockwise

One way of doing this easily is to use the following 'quick' formula:

Moment = $Fd\sin\theta$ F = magnitude of force d = distance

θ = angle between force and line distance is measured along

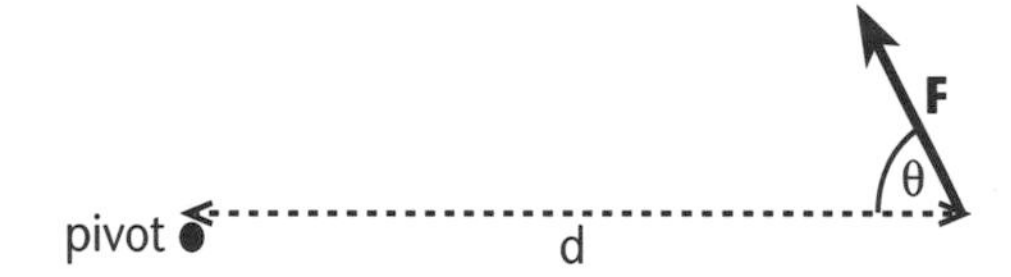

If you decide to use this formula, you MUST make sure you use the correct angle for θ – you may need to use angle facts to work it out

Using Moments to Solve Problems

If a body is in equilibrium, total clockwise moments = total anticlockwise moments about any point.

We can use this fact, together with resolving (see *AS in a Week* if you are not happy about this!) to find unknown forces:

Moments

Example 1

A uniform rod of length 2 m and weight 10 N rests on supports 50 cm from each end. Find the magnitude of the reaction force at the supports.

"Uniform" means the centre of mass is in the middle

Do NOT assume the reaction is the same at both pivots!

Solution

Step 1 Draw a diagram, showing all forces and distances

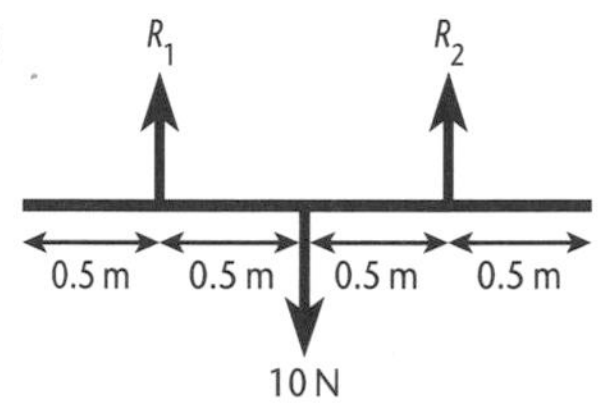

Step 2 Choose a point to take moments about – it should be the point an unknown force goes through

Take moments about left-hand support

Step 3 Use total clockwise moments = total anticlockwise moments

$10 \times 0.5 = R_2 \times 1$
$R_2 = 5\text{N}$

Step 4 Either resolve or use moments about another point to find any remaining forces

Resolving vertically:
$R_1 + R_2 = 10$
$R_1 = 5\text{N}$

If you are not sure which forces give clockwise moments and which give anticlockwise moments, try using a ruler on the table as the rod, a finger as the pivot and push it in the direction of the force – you can then see which way it turns!

Example 2

AB is a non-uniform plank of length $5a$ and mass M. It is supported at points C and D, where AC = a and AD = $3a$.

A particle of mass M is placed at B. The plank is now about to lose contact with the support at C. Find the distance of the centre of mass of the plank from A.

Solution

1.

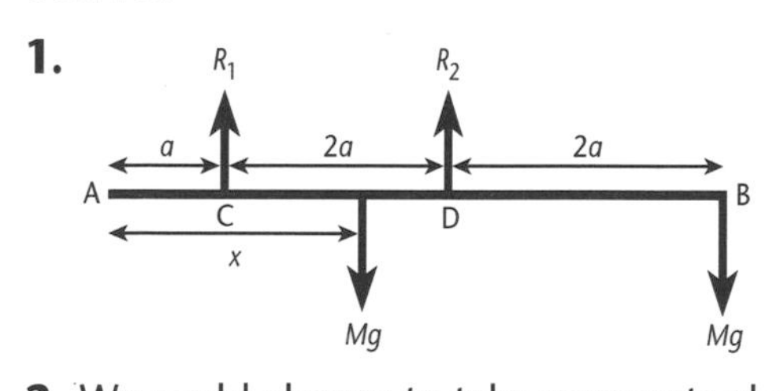

Since we don't know the distance of the centre of mass from A, we call it x

Remember this!

2. We could choose to take moments about C or D, since they both have unknown forces acting. But, the plank is about to lose contact at C, so $R_1 = 0$
So we'll take moments about D

3. $(3a - x) \times Mg = 2a \times Mg \qquad 3a - x = 2a \qquad x = a$

Since AD = 3a, and centre of mass is x from A, distance of centre of mass from D is 3a − x

Use your knowledge

1 A uniform plank AB of mass 10 kg is 5 m long. It is supported at points C and D, where AC = 1 m and AD = 3 m.

a) Find the magnitude of the reactions at points C and D, giving your answer in terms of g.

The reactions are not the same.
Take moments about either C or D.
Resolve to find the other reaction force

A cat walks along the plank from end A towards end B. The plank starts to tilt when the cat is 50 cm from B.

b) Find the mass of the cat.

The reaction at C will become zero
Take moments about D

c) State two modelling assumptions that you have made in answering this question.

In mechanics, you model things as a "particle", a "rod" or a "lamina". Words such as "light" (= no mass) or "smooth" (= no friction) are also modelling assumptions – but if they're given in the question, you cannot quote them!

2 A uniform rod AB of length $6a$ and weight 20 N rests with end A on rough, horizontal ground and a smooth support at point C, where BC = a. The rod makes an angle of 60° with the ground.

Find

a) The magnitude of the reaction force at the support at point C.

Take moments about A
You need only a component of the weight
Reaction at C is perpendicular to the rod

b) The coefficient of friction between the rod and the ground, given that the rod is about to slip. Give your answer in surd form.

Resolve horizontally and vertically
Use $F = \mu R$

Particle Kinematics

Test your knowledge

1 The displacement, s (in metres), of a particle P, moving in a straight line relative to a fixed point O, after t seconds is given by the formula:

$s = t^3 - 6t^2 - 15t + 2; t \geq 0$

a) Find the speed and acceleration of the particle P, when $t = 1$ s.
b) Find the time when the particle is instantaneously at rest.

2 A particle P, moves in a straight line along the Ox axis. Initially the particle is moving with velocity $24\,\text{ms}^{-1}$ in the direction of x increasing. The acceleration of P, t seconds after leaving the origin, is given by the formula: $a = 14 - 6t\,\text{ms}^{-2}$

a) Find expressions for the particle's velocity, v, and displacement, s, in terms of t
b) Calculate the time at which the particle P is instantaneously at rest
c) Sketch a velocity time graph for the particle P in the time $0 \leq t \leq 8$
d) Find the displacement of the particle 8 seconds after initially leaving the origin
e) Find the distance travelled by the particle in the time interval $0 \leq t \leq 8$

3 A particle P moves in a straight line. Its velocity, at time t seconds after passing through the origin, O, is given by the formula: $v = 2t^2 + t - 1$. Find:

a) the acceleration of P, when its velocity is $0\,\text{ms}^{-1}$
b) the time, to 2 significant figures, at which the particle returns to the origin, O.

4 The position vector of a particle P, of mass 1.5 kg, moving in an horizontal plane at time t seconds, relative to the origin, O, is given by:

$\mathbf{r} = (t^2 - 3t)\mathbf{i} + (t - 5)\mathbf{j}$ m, $t \geq 0$ where $\mathbf{i}$ and $\mathbf{j}$ are unit vectors due east and due north respectively. At the instant when $t = 4$ s, find:

a) the speed (in ms^{-1}) of the particle P
b) the magnitude and direction of the force acting on P.

Answers

1 a) $24\,\text{ms}^{-1}$, $-6\,\text{ms}^{-2}$ **b)** 5 s
2 a) $v = 14t - 3t^2 + 24$, $a = 7t^2 - t^3 + 24t$
b) $t = 6$ s **c)** see diagram **d)** 128 m
e) 232 m **3 a)** $3\,\text{ms}^{-2}$ **b)** 0.91 s
4 a) $5.10\,\text{ms}^{-1}$ **b)** 3 N due east

v (ms^{-1})
24
t (s)
6
8
−56

If you got them all right, skip to page 100

Particle Kinematics

Improve your knowledge

Key points from AS in a Week

Vectors in Mechanics pages 59–65

There are many circumstances in mechanics where the acceleration of a particle, moving in a straight line is not constant. In these situations we cannot use the equations of motion such as:

$v = u + at, \qquad s = ut + \frac{1}{2}at^2, \qquad v^2 = u^2 + 2as, \qquad s = \frac{1}{2}(u + v)t$

This is because these equations only work when the acceleration is constant.

When the acceleration of a body moving in a straight line is not constant (i.e. variable) we must use differentiation or integration to help us to find equations for displacement, velocity and acceleration in terms of the time, t.

1 Use of Differentiation

$v = \frac{ds}{dt}$ because the velocity, v, is the rate of change of displacement, s

$a = \frac{dv}{dt}$ because the acceleration, a, is the rate of change of velocity, v

Example 1

The displacement, s (in m), of a particle P, relative to a fixed point O, after t seconds, is given by the formula: $s = 2t^3 - 3t + 2$. The particle moves in a straight line.

a) Find the velocity and acceleration of the particle P, when $t = 3$ s
b) Find the time when the particle is instantaneously at rest.

Solution

a) $v = \frac{ds}{dt} = 6t^2 - 3$ Hence, when $t = 3$, $v = 6(3^2) - 3 = 51\,\text{ms}^{-1}$

$a = \frac{dv}{dt} = 12t$ Hence, when $t = 3$, $a = 12(3) = 36\,\text{ms}^{-2}$

b) instantaneously at rest $\Rightarrow v = 0 \Rightarrow 0 = 6t^2 - 3 \Rightarrow t^2 = \frac{3}{6} \Rightarrow t = 0.707$ s

2 Use of Integration

When acceleration is expressed in terms of time t, we can use integration with respect to t, to form the following relationships:

Velocity, $v = \int a\,dt$ and displacement, $s = \int v\,dt$

When you integrate, you must always remember to include a constant of integration. Usually you will be given boundary conditions to enable you to calculate the constant of integration.

Example 2

A particle P, moves in a straight line along the Ox axis. Initially the particle is moving with velocity $8\,ms^{-1}$ in the direction of x decreasing. The acceleration of P, t seconds after leaving the origin, is given by the formula: $a = 6t - 10\,ms^{-2}$

a) Find expressions for the particle's velocity, v, and displacement s, in terms of t
b) Calculate the time at which the particle P is instantaneously at rest
c) Sketch a velocity–time graph for the particle P in the time interval $0 \le t \le 6$
d) Find the displacement of the particle 6 seconds after initially leaving the origin
e) Find the distance travelled by the particle in the time interval $0 \le t \le 6$

Solution

- Write down boundary conditions (BCs): $t = 0\,s \Rightarrow s = 0\,m,\ v = -8\,ms^{-1}$

a) Integrate a to obtain v: $= \int 6t - 10\,dt = 3t^2 - 10t + c$

- Use BCs to evaluate c: $t = 0,\ v = -8 \Rightarrow -8 = 3(0)^2 - 10(0) + c \Rightarrow c = -8$
- Hence: velocity, $v = 3t^2 - 10t - 8$
- Integrate v to obtain s: $s = \int 3t^2 - 10t - 8\,dt = t^3 - 5t^2 - 8t + k$
- Use BCs to evaluate k: $t = 0,\ s = 0 \Rightarrow 0 = (0)^3 - 5(0)^2 - 8(0) + k \Rightarrow k = 0$
- Hence: displacement, $s = t^3 - 5t^2 - 8t$

b) Instantaneously at rest $\Rightarrow v = 0 \Rightarrow 0 = 3t^2 - 10t - 8$

- Solve for t: $(3t + 2)(t - 4) = 0 \Rightarrow t = -\frac{2}{3}$ (reject) and $t = 4\,s$ (accept!)

c) $v = 3t^2 - 10t - 8 = (3t + 2)(t - 4)$ is a quadratic with roots at $t = -\frac{2}{3}$ and $t = 4$

- When $t = 0$, $v = 2 \times -4 = -8\,ms^{-1}$
- When $t = 6$, $v = 20 \times 2 = 40\,ms^{-1}$

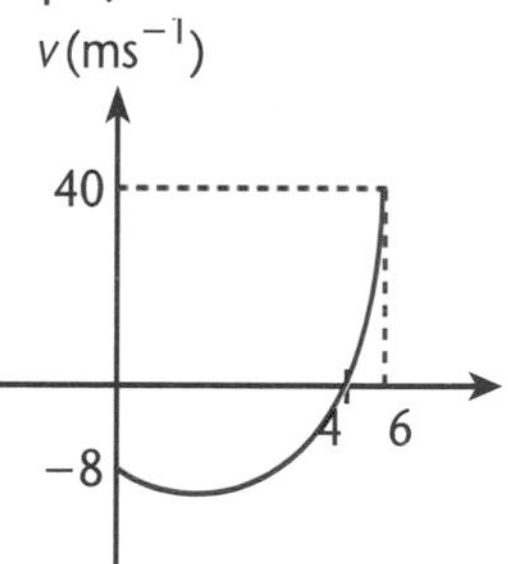

d) Sub. $t = 6\,s$ into s: $s = 216 - 5(36) - 8(6) = -12\,m$

e) Looking at the velocity–time graph above, we see that the velocity is negative (below the t-axis) between $t = 0$ and 4 seconds. During this time the particle, P, is moving to the left (or in a negative x-direction).

- At 4 seconds the particle stops and is instantaneously at rest (i.e. $v = 0\,ms^{-1}$)
- However after 4 seconds, the particle's velocity is positive (above the t-axis), which means that the particle then moves to the right
- Sub. $t = 4\,s$ into s: $s = 64 - 5(16) - 8(4) = -48\,m$
- From diagram opposite: Distance $= 48 + 36 = 84\,m$

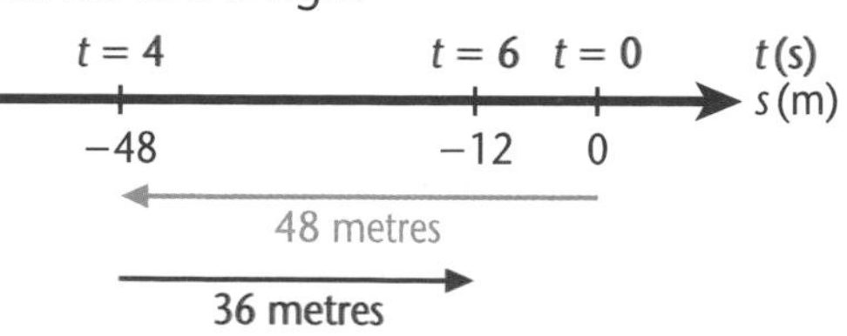

3 Use of Differentiation and Integration

Some questions on particle kinematics will require you to use a mixture of differentiation and integration. The diagram below helps you to decide what to do:

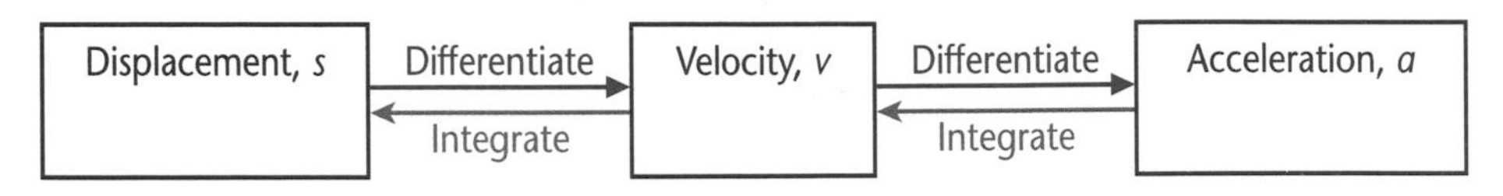

Example 3

The velocity of a particle P, at time t seconds after passing through the origin, O is given by the formula: $v = 6t - t^2 + 12\,\text{ms}^{-1}$. Find:

a) the acceleration of the particle P, when its velocity is $21\,\text{ms}^{-1}$
b) the time at which the particle, P returns to the origin, O, given that the particle moves in a straight line.

Solution

a) Find t, when $v = 21$: $21 = 6t - t^2 + 12 \Rightarrow t^2 - 6t + 9 = 0$
- Factorise and solve: $(t - 3)(t - 3) = 0 \Rightarrow t = 3\,\text{s}$
- Differentiate from v to a: $a = \dfrac{dv}{dt} = 6 - 2t$
- When $t = 3\,\text{s}$: $a = 6 - 2(3) = 0\,\text{ms}^{-2}$

b) When the particle returns to the origin, O, the displacement, $s = 0\,\text{m}$
- So we integrate from v to s: $s = \int 6t - t^2 + 12\,dt = 3t^2 - \frac{t^3}{3} + 12t + c$
- Write down BCs: When $t = 0$, $s = 0$
- Use BCs to evaluate c: $0 = 0 - 0 + 0 + c \Rightarrow c = 0$
- Hence: displacement, $s = 3t^2 - \frac{t^3}{3} + 12t$
- Apply $s = 0\,\text{m}$: $0 = 3t^2 - \frac{t^3}{3} + 12t = -\frac{t}{3}(-9t + t^2 - 36) = -\frac{t}{3}(t^2 - 9t - 36)$
- Factorise and solve: $0 = -\frac{t}{3}(t - 12)(t + 3) \Rightarrow t = 0, 12, -3 \Rightarrow t = 12\,\text{s}$

4 Use of Vectors

Particles do not always move in straight lines. Their motion may be either two- or three-dimensional. To express motion in these scenarios, we use vectors, which are dependent on time:

r: position vector telling us where the particle is relative to a fixed origin O
v: the velocity vector
a: the acceleration vector

Particle Kinematics

When we use vectors to describe a particle's motion, the relationships between displacement, velocity and acceleration apply in the same way:

Differentiation: velocity, $\mathbf{v} = \dfrac{d\mathbf{s}}{dt}$ acceleration, $\mathbf{a} = \dfrac{d\mathbf{v}}{dt}$

Integration: position vector, $\mathbf{r} = \int \mathbf{v}\, dt$ velocity, $\mathbf{v} = \int \mathbf{a}\, dt$

Also, we may be asked to find the scalar equivalent of a vector quantity. To do this we work out the magnitude of the vector using Pythagoras.

Remember: The magnitude of the velocity vector $\mathbf{v}$, is the speed of the particle. The magnitude of the position vector, $\mathbf{r}$, is the distance of a particle from the origin.

Example 4

A particle P of mass 0.8 kg moves with a velocity of $\mathbf{v} = (3t^2 + 2)\mathbf{i} + (2t - 3)\mathbf{j}\,\text{ms}^{-1}$ When $t = 0\,\text{s}$, the displacement of P from the origin O, is given by the position vector, $\mathbf{r} = 3\mathbf{i} - 2\mathbf{j}\,\text{m}$. Find, to 3 significant figures, when $t = 2\,\text{s}$:

a) the speed of the particle P
b) the magnitude of the force acting on the particle
c) the distance of P from the origin.

Solution

a) When $t = 2\,\text{s}$: $\mathbf{v} = (12 + 2)\mathbf{i} + (4 - 3)\mathbf{j} = 14\mathbf{i} + 1\mathbf{j}$

Speed $\Rightarrow$ Use Pythagoras on $\mathbf{v}$: Speed $= \sqrt{(14)^2 + (1)^2} = 14.0\,\text{ms}^{-1}$ (3sf)

b) To find the force we use: $\mathbf{F} = m\mathbf{a}$
Given: $\mathbf{v} = (3t^2 + 2)\mathbf{i} + (2t - 3)\mathbf{j} \Rightarrow \mathbf{a} = \dfrac{d\mathbf{v}}{dt} = 6t\,\mathbf{i} + 2\mathbf{j}$

Apply $\mathbf{F} = m\mathbf{a}$: $\mathbf{F} = 0.8(6t\,\mathbf{i} + 2\mathbf{j}) = 4.8t\mathbf{i} + 1.6\mathbf{j}$
When $t = 2\,\text{s} \Rightarrow \mathbf{F} = 9.6\mathbf{i} + 1.6\mathbf{j}\,\text{N}$

Magnitude $\Rightarrow$ Use Pythagoras: $|\mathbf{F}| = \sqrt{(9.6)^2 + (1.6)^2} = 9.73\,\text{N}$ (3sf)

c) Integrate from $\mathbf{v}$ to $\mathbf{r}$: $\mathbf{r} = \int \mathbf{v}\, dt = \int (3t^2 + 2)\mathbf{i} + (2t - 3)\mathbf{j}\, dt$
Hence: $\mathbf{r} = (t^3 + 2t)\mathbf{i} + (t^2 - 3t)\mathbf{j} + \mathbf{c}$, where $\mathbf{c}$ is the vector constant of integration.

BCs: $\mathbf{r} = 3\mathbf{i} - 2\mathbf{j}\,\text{m}$, when $t = 0\,\text{s} \Rightarrow 3\mathbf{i} - 2\mathbf{j} = 0\mathbf{i} + 0\mathbf{j} + \mathbf{c} \Rightarrow \mathbf{c} = 3\mathbf{i} - 2\mathbf{j}$

Hence, position vector: $\mathbf{r} = (t^3 + 2t + 3)\mathbf{i} + (t^2 - 3t - 2)\mathbf{j}\,\text{m}$

When $t = 2\,\text{s}$: $\mathbf{r} = (8 + 4 + 3)\mathbf{i} + (4 - 6 - 2)\mathbf{j} = 15\mathbf{i} - 4\mathbf{j}\,\text{m}$

Distance from 0 $\Rightarrow$ magnitude of $\mathbf{r}$: $|\mathbf{r}| = \sqrt{(15)^2 + (-4)^2} = 15.5\,\text{m}$ (3sf)

Particle Kinematics

45 minutes

Use your knowledge

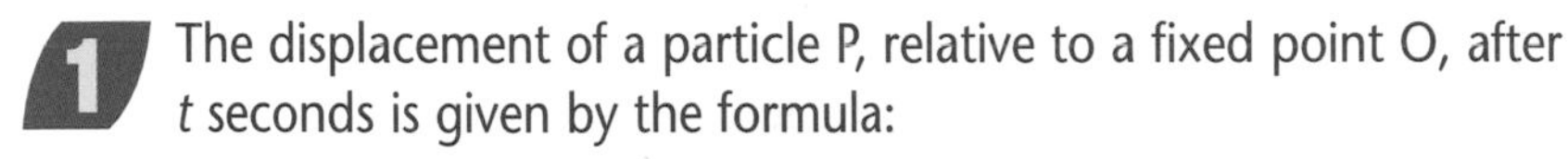

1 The displacement of a particle P, relative to a fixed point O, after t seconds is given by the formula:

$s = t^3 - 7t^2 + 8t + 4\,\text{m};\ t \geq 0$

Given that the particle moves in a straight line,

a) Find the times at which the particle P is instantaneously at rest.

When $v = 0\,\text{ms}^{-1}$. Factorise and solve.

b) Hence, find the distance travelled, to the nearest metre, by the particle in the time interval $0 \leq t \leq 6$

Find s, when $t = 0\,s$ & $6\,s$ and for values of t from part a)

2 The acceleration (in ms^{-2}) of a particle P, at time t seconds, is given by the formula:

$\mathbf{a} = 3\mathbf{i} + (2t - 1)\mathbf{j}\ ;\ t \geq 0$

where $\mathbf{i}$ and $\mathbf{j}$ are unit vectors due east and north respectively.
When $t = 1\,\text{s}$, the velocity of P is $27\mathbf{i} + 6\mathbf{j}$.

Boundary conditions

a) Find the velocity vector, $\mathbf{v}$, of P, at time t seconds.

Integrate and use BCs.

b) Determine the time(s) if any, when the particle P is moving parallel to the vector $2\mathbf{i} + \mathbf{j}$

Look at the velocity vector. Ratio of components $\mathbf{i}$:$\mathbf{j}$ is 2:1

Given that the particle P is initially at the origin O, find:

Boundary condition.

c) the distance OP, to 3 significant figures, when $t = 1\,\text{s}$

OP is the magnitude of $\mathbf{r}$

3 The resultant force $\mathbf{F}$ acting on a particle P, when $t \geq 0$, is given by:
$\mathbf{F} = (8t - 3)\mathbf{i} + 2t\mathbf{j}\,\text{N}$
When $t = 0\,\text{s}$, the particle of mass 4 kg moves through the origin with velocity $-\frac{5}{2}\mathbf{i} - \mathbf{j}\,\text{ms}^{-1}$.

Boundary conditions

a) Find the velocity vector, $\mathbf{v}$, of P in terms of t

Use $\mathbf{F} = m\mathbf{a}$ to find $\mathbf{v}$.

b) Find the time(s), if any for which the particle P, is instantaneously at rest.

Both $\mathbf{i}$ and $\mathbf{j}$ components of $\mathbf{v}$, must be zero at the same time.

Given that the points A and B, are the positions of the particle P, when $t = 1\,\text{s}$ and $t = 2\,\text{s}$, respectively,

Find $\mathbf{r}$, at $t = 1s$ and $t = 3s$

c) calculate, to 3 significant figures, the distance AB.

Use distance between two points formula from Pure Maths

Binomial and Poisson Distributions

Test your knowledge

(Cumulative binomial and Poisson tables must be used for some questions)

1

i) Research has shown that the probability that a British person has blood group A is 0.42. Calculate, to 3 significant figures, the following probabilities, for a sample of eight British people, chosen at random:

a) two have blood group A   b) at most three have blood group A

c) at least four have blood group A.

ii) Let $X \sim \text{Bin}(10, 0.4)$. Use your cumulative binomial tables to find the following to 4 decimal places:

a) $P(X \geq 5)$ b) $P(X < 7)$ c) $P(2 \leq X \leq 6)$

d) $P(X > 8)$ e) $P(4 < X < 8)$ f) $P(3 < X \leq 8)$

iii) Let $X \sim \text{Bin}(8, 0.27)$. Find, to 3 significant figures:

a) μ b) σ c) $P(\mu - \sigma < X < \mu)$

2

The number of car thefts reported to Wandsell Police Station averages 3.75 per week. Assuming that the number of car thefts fit a Poisson distribution, find the following probabilities:

a) exactly three car thefts are reported in a week

b) at most three car thefts are reported in a week

c) fewer than ten car thefts are reported in a two-week interval.

3

a) Using a suitable approximation, find the probability that you get between 55 and 65 heads, inclusive, when you toss a fair coin 120 times.

b) Research has shown that the probability that a British person has blood group AB is 0.03. A sample of 150 British people is chosen at random. Find, using a suitable approximation, the probability that at least six of them have blood group AB.

c) The local shop sells packets of 'Supa Nappies' at a rate of 3.2 per week. It is assumed that these packets sell independently of each other and follow a Poisson distribution. Calculate, using a suitable approximation, the probability that over a 52-week year, the local shop sells more than 180 packets of Supa Nappies.

Answers

1 i) a) 0.188 **b)** 0.547 **c)** 0.453 **ii) a)** 0.3669 **b)** 0.9452 **c)** 0.8988 **d)** 0.0017 **e)** 0.3546 **f)** 0.6160 **iii) a)** 2.16 **b)** 1.26 **c)** 0.548 **2 a)** 0.207 **b)** 0.484 **c)** 0.776 **3 a)** 0.683 to 0.685 **b)** 0.297 **c)** 0.137 to 0.138

If you got them all right, skip to page 108

Binomial and Poisson Distributions

Improve your knowledge

1 Binomial Distribution

If the random variable X is defined as the number of successful trials and when the following conditions are true:

a) there are a fixed number of trials, n
b) each trial consists of two outcomes – success or failure
c) the probability of success per trial, p, is constant
d) the trials are independent of each other

then X has a binomial distribution with index n and parameter p, i.e. $X \sim \text{Bin}(n, p)$ and $P(X = x) = {}^nC_x p^x(1 - p)^{n-x}$ for $x = 0,1,2,\ldots,n$

Also, for a binomial distribution $E(X) = np$ and $\text{Var}(X) = np(1 - p)$

Example 1

A fair cubical die is thrown 10 times and the score on the die is recorded. Find the probability that:

a) 2 sixes are thrown, b) at most 3 sixes are thrown, c) at least 2 sixes are thrown.

State the mean number of sixes that are thrown.

Solution

Define X = number of sixes that are thrown.

At most means ≤
At least means ≥

$X \sim \text{Bin}(10, \frac{1}{6})$

a) $P(X = 2) = {}^{10}C_2\left(\frac{1}{6}\right)^2\left(\frac{5}{6}\right)^8 = 0.29071\ldots = 0.291$ (3*sf*)

b) $P(X \leq 3) = P(X = 0) + P(X = 1) + P(X = 2) + P(X = 3)$

$$= {}^{10}C_0\left(\frac{1}{6}\right)^0\left(\frac{5}{6}\right)^{10} + {}^{10}C_1\left(\frac{1}{6}\right)^1\left(\frac{5}{6}\right)^9 + 0.29071\ldots + {}^{10}C_3\left(\frac{1}{6}\right)^3\left(\frac{5}{6}\right)^7$$
$$= 0.16151\ldots + 0.32301\ldots + 0.29071\ldots + 0.15505\ldots$$
$$= 0.93028\ldots = 0.930 \text{ (3sf)}$$

c) $P(X \geq 2) = 1 - P(X \leq 1) = 1 - (P(X = 0) + P(X = 1))$
$$= 1 - (0.16151\ldots + 0.32301\ldots)$$
$$= 0.51548\ldots = 0.515 \text{ (3sf)}$$

Mean number of sixes $\Rightarrow E(X) = np = 10\left(\frac{1}{6}\right) = 1.67$ (3sf)

Use of the Cumulative Binomial Tables

In your formula book you are given cumulative binomial tables. These give you $P(X \le x)$ for certain values of n and for p.

Example 2

Let $X \sim \text{Bin}(10, 0.3)$. Use your cumulative tables to find to 4 decimal places:

i) a) $P(X \le 4)$ b) $P(X < 6)$ c) $P(X \ge 7)$
d) $P(X > 3)$ e) $P(X = 6)$ f) $P(X = 0)$
g) $P(2 \le X \le 7)$ h) $P(1 < X \le 6)$ i) $P(5 \le X < 8)$

ii) Find μ and σ.
iii) Work out $P(\mu - \sigma < X < \mu + \sigma)$ to 3 significant figures.

Solution

i) a) $P(X \le 4) = 0.8497$ (from tables)
b) $P(X < 6) = P(X \le 5) = 0.9527$ (from tables)
c) $P(X \ge 7) = 1 - P(X \le 6) = 1 - 0.9894 = 0.0106$
d) $P(X > 3) = 1 - P(X \le 3) = 1 - 0.6496 = 0.3504$
e) $P(X = 6) = P(X \le 6) - P(X \le 5)$
$= 0.9894 - 0.9527 = 0.0367$
f) $P(X = 0) = 0.0282$ (from tables)
g) $P(2 \le X \le 7) = P(X \le 7) - P(X \le 1)$
$= 0.9984 - 0.1493 = 0.8491$
h) $P(1 < X \le 6) = P(X \le 6) - P(X \le 1)$
$= 0.9894 - 0.1493 = 0.8401$
i) $P(5 \le X < 8) = P(X \le 7) - P(X \le 4)$
$= 0.9984 - 0.8497 = 0.1487$

Binomial Tables
n = 10, *p* = 0.3

$x =$	$P(X \le x)$
0	0.0282
1	0.1493
2	0.3828
3	0.6496
4	0.8497
5	0.9527
6	0.9894
7	0.9984
8	0.9999
9	1.0000

ii) $\mu = E(X) = np = 10 \times 0.3 = 3$
$\sigma^2 = \text{Var}(X) = np(1 - p) = 10 \times 0.3 \times 0.7 = 2.1$
$\Rightarrow \sigma = \sqrt{2.1} = 1.449... = 1.45$ (3sf)

iii) $P(\mu - \sigma < X < \mu + \sigma) = P(3 - 1.45 < X < 3 + 1.45) = P(1.55 < X < 4.45)$
$= P(2 \le X \le 4)$ (since $X \sim$Bin, a discrete distribution)
$= P(X \le 4) - P(X \le 1) = 0.8497 - 0.1493 = 0.7004$
$= 0.700$ (3sf)

Binomial and Poisson Distributions

2 Poisson Distribution

If X is defined as 'the number of events' and when the following conditions are true:

a) each event occurs independently of another
b) events occur singly and randomly (and are not clustered)
c) events occur at a constant rate, which is proportional to time or space

then X has a Poisson distribution with mean rate (parameter) λ, i.e. $X \sim \text{Poi}(\lambda)$

and $P(X = x) = \dfrac{e^{-\lambda}\lambda^x}{x!}$ for $x = 0, 1, 2, \ldots$

Also, for a Poisson distribution $E(X) = \lambda$ and $Var(X) = \lambda$

Remember: A Poisson distribution has the same mean and variance!

Exam questions will test your use of the Poisson formula or Poisson cumulative tables, which are found in your formula book.

Example 3

The number of flaws per 10 m^2 of cloth produced by a factory is modelled by a Poi(2.25) distribution.

a) Find the probability that for a randomly selected 10 m^2 piece of cloth, there are:
   i) no flaws, ii) two flaws, iii) more than three flaws.

b) Calculate the probability that there are more than 12 flaws in a randomly selected 8 m by 5 m piece of cloth.

   Khallidur, the quality control manager, selects 8 pieces of 10 m^2 cloth, at random, from the factory's output. He decides to order new machinery to improve the quality of cloth if he finds at least two pieces of cloth with more than three flaws in them.

c) Calculate the probability that Khallidur orders new machinery.

Solution

Define X = number of flaws

a) For 10 m^2 ; $X \sim \text{Poi}(2.25)$

   i) $P(X = 0) = \dfrac{e^{-2.25}(2.25)^0}{0!} = e^{-2.25} = 0.10540\ldots = 0.105\ (3sf)$

   ii) $P(X = 2) = \dfrac{e^{-2.25}(2.25)^2}{2!} = 0.26679\ldots = 0.267\ (3sf)$

iii) $P(X > 3) = 1 - (P(X = 0) + P(X = 1) + P(X = 2) + P(X = 3))$

$$= 1 - \left(0.10540... + \frac{e^{-2.25}(2.25)^1}{1!} + 0.26679... + \frac{e^{-2.25}(2.25)^3}{3!}\right)$$

$$= 1 - (0.10540... + 0.23715... + 0.26679... + 0.20009...)$$
$$= 0.19057...$$

$$= 0.191 \text{ (3}sf\text{)}$$

b) 8 m by 5 m $\Rightarrow$ 40 m^2 of cloth

- New distribution $\Rightarrow X \sim \text{Poi}(4 \times 2.25) = \text{Poi}(9)$ for 40 m^2.
- Hence: $P(X > 12) = 1 - P(X \le 12) = 1 - 0.8758$ (from cumulative tables)
  $= 0.1242 = 0.124$ (3sf)

c) This is an example of a 'secret binomial'.

- Define Y = number of 10 m^2 pieces of cloth with more than 3 flaws
- Hence $Y \sim \text{Bin}(8, 0.19057...)$ using answer to a) iii)
  $P(Y \ge 2) = 1 - P(Y \le 1) = 1 - (P(Y = 0) + P(Y = 1))$
  $= 1 - \left({}^8C_0(0.19057)^0(0.80943)^8 + {}^8C_1(0.19057)^1(0.80943)^7\right)$
  $= 1 - (0.18426... + 0.34706...) = 0.46868... = 0.469$ (3sf)

There are a fixed, n = 8 pieces of 10 m^2 cloth. p = 0.19057, is the probability that one 10 m^2 cloth contains more than 3 flaws

3 Approximations

There are three main approximations that you need to know in A-Level Statistics. They are (with conditions):

a) $X \sim \text{Bin}(n, p)$ (discrete) $\xrightarrow[n > 20,\ p \le 0.1]{\text{approx.}}$ $X \sim \text{Poi}(np)$ (discrete)

b) $X \sim \text{Bin}(n, p)$ (discrete) $\xrightarrow[n > 5,\ np(1 - p) > 5]{\text{approx.}}$ $X \sim N(np, np(1 - p))$ (continuous)

c) $X \sim \text{Poi}(\lambda)$ (discrete) $\xrightarrow[\lambda > 10]{\text{approx.}}$ $X \sim N(\lambda, \lambda)$ (continuous)

The conditions given vary from teacher to teacher, exam board to exam board. So please consult your exam board about the specific conditions they require.

Binomial and Poisson Distributions

Students find difficulty in choosing the correct approximation. If you start with a Poisson distribution you know you must go to a normal (if λ is big). If you start with a binomial (with big n), first check is p is small (i.e. < 0.1); if it is, then approximate to a Poisson; if not, then approximate to a normal and then check to see if the value of np and $np(1 - p)$ are bigger than 5.

Continuity Corrections

When using the Bin → Normal, or Poi → Normal approximations, you must apply a continuity ($\pm \frac{1}{2}$) correction. You apply a continuity correction (c.c.) because this takes into account that you are approximating a discrete distribution with a continuous normal distribution.

The table below gives you a general rule for applying continuity corrections (although there are other ways to work them out).

Type of probability	What do you do to a
$P(X \geq a)$	$-\frac{1}{2}$
$P(X > a)$	$+\frac{1}{2}$
$P(X \leq a)$	$+\frac{1}{2}$
$P(X < a)$	$-\frac{1}{2}$

It is very important that you apply the continuity correction correctly. Here are some examples:

a) $P(X \geq 12) \cong P(X > 11.5)$
b) $P(X \leq 5) \cong P(X < 5.5)$
c) $P(X > 15) \cong P(X > 15.5)$
d) $P(X < 7) \cong P(X < 6.5)$
e) $P(10 \leq X \leq 18) \cong P(9.5 < X < 18.5)$
f) $P(5 < X \leq 20) \cong P(5.5 < X < 20.5)$
g) $P(12 \leq X < 28) \cong P(11.5 < X < 27.5)$
h) $P(4 < X < 10) \cong P(4.5 < X < 9.5)$

Example 4

Sweenies make coloured sweets that are packed at random, with each packet containing 100 sweets. It is known over a long period of time that the proportions of green and blue coloured sweets packed are 0.23 and 0.04 respectively.

If X is defined as 'the number of green sweets in a packet of Sweenies':

i) Explain why X could be modelled by a binomial distribution.

ii) Find (using suitable approximations) the probability that a packet contains:

a) between 20 and 30 green sweets, inclusive

b) fewer than four blue sweets.

Solution

Make reasons relevant to question

i)
- There are a fixed number of sweets in a packet of Sweenies,
- there is a constant probability (0.23) that the sweet is green
- a sweet can either be green or not green
- colours of each sweet examined are independent of each other.

ii) a) Distribution: $X \sim \text{Bin}(100, 0.23) \xrightarrow{\text{approx.}} X \sim \text{N}(23, 17.71)$

Need: $P(20 \le X \le 30) \cong P(19.5 < X < 30.5)$, using a continuity correction

z-values: $\dfrac{19.5 - 23}{\sqrt{17.71}} = -0.832\ldots \qquad \dfrac{30.5 - 23}{\sqrt{17.71}} = 1.782\ldots$

$= P(-0.83 < z < 1.78) = \Phi(0.83) + \Phi(1.78) - 1 = 0.7967 + 0.9625 - 1$
$= 0.7592 = 0.759$ (3sf) (0.760 by interpolation)

b) Define: Y = number of blue sweets in a packet of Sweenies

- Distribution: $Y \sim \text{Bin}(100, 0.04) \xrightarrow{\text{approx.}} Y \sim \text{Poi}(4)$
- Need: $P(Y < 4) = P(Y \le 3) = 0.4335 = 0.434$ (3sf) from tables.

Example 5

A radioactive source emits radioactive particles at a rate of 0.4 per minute. Assuming that the number of radioactive particles emitted fits a Poisson distribution, calculate, by using a suitable approximation, the probability that this source emits at least 55 radioactive particles in two hours.

Solution

Define: X = number of radioactive particles emitted
$X \sim \text{Poi}(0.4)$ per minute $\Rightarrow X \sim \text{Poi}(120 \times 0.4) = \text{Poi}(48)$ for two hours

Since λ is large; $X \sim \text{Poi}(48) \xrightarrow{\text{approx.}} X \sim \text{N}(48, 48)$
Need: $P(X \ge 55) \cong P(X > 54.5)$ (using a continuity correction)

z-value: $\dfrac{54.5 - 48}{\sqrt{48}} = 0.938\ldots$

$P(z > 0.94) = 1 - \Phi(0.94) = 1 - 0.8264 = 0.1736 = 0.174$ (3sf) (0.174 by interpolation)

Binomial and Poisson Distributions

Use your knowledge

1 The number of telephone calls to a switchboard of a busy accountancy practice is known to follow a Poisson distribution with an average rate of 4.75 calls every 15 minutes. Find the probability that there are:

a) exactly 6 calls in a 15-minute period — *Use the Poisson formula!!*

b) at least 10 calls in a 30-minute period — *New distribution for 30 mins!!*

c) more than 50 calls in a 3-hour period — *What do you do if λ is big?*

d) exactly 6 calls in each of two successive 15-minute periods. — *Use your answer to part (a)*

2 Customers at Wandsell Supermarket can pay for their goods by either cash, debit card or cheque. Over a long period of time it is known that the ratio of customers paying by either cash, debit card or cheque is 11:8:1. — *What is the distribution? You need tables for (a) to (c)*

i) Calculate the following probabilities for a random sample of 25 customers:

a) more than 10 pay by debit card — *Do 1 – Probs*

b) fewer than 14 pay by cash — *p = 0.55 is not in your tables.*

c) more than 12 do not pay by cash — *What 'p' do not pay by cash?*

d) the expected number who pay by debit card. — *What is the mean of a binomial?*

ii) During one evening 80 customers purchase goods at Wandsell supermarket. Using a suitable approximation, find the probability that:

a) fewer than 5 customers pay by cheque — *n is big and p is small.*

b) between 48 and 53 customers inclusive pay by cash. — *n is big, but p is not small. c.c.!*

3 The number of birthday cakes sold at Beryl's shop is known to follow a Poisson distribution with a mean of 6 per day.

i) Find the probability that during a particular day:

a) 4 birthday cakes are sold — *Use formula or tables*

b) at least 6 birthday cakes are sold. — *Use your tables to help*

ii) Beryl's shop is open for 5 days every week. What is the probability that on exactly 3 days, in a randomly chosen week, Beryl sells at least 6 birthday cakes? — *Beware, the secret binomial!!*

Normal Distribution

Test your knowledge

$X \sim N(5, 4)$. Find

a) $P(X < 6)$ b) $P(X > 6.5)$ c) $P(X < 2)$
d) $P(X > 4.5)$ e) $P(6 < X < 7.2)$ f) $P(4.3 < X < 5.1)$

$X \sim N(4, 3)$. Find the values of a, b, c and d, given that:

a) $P(X > a) = 0.1$ b) $P(X < b) = 0.99$
c) $P(X > c) = 0.95$ d) $P(X < d) = 0.025$

The speeds of cars passing a certain point are known to follow a normal distribution. 20% of cars travel at 50 mph or above, and 10% of cars travel at 25 mph or below. Find the mean and standard deviation of the speeds of the cars.

A fair die is rolled 216 times, and the number of sixes recorded.

a) State the mean and variance of the number of sixes recorded.
b) Use a suitable approximation to determine the probability that at least 40 sixes were recorded.

Answers

1 a) 0.6915 **b)** 0.2266 **c)** 0.0668 **d)** 0.5987 **e)** 0.1728 **f)** 0.1567
2 a) 6.22 **b)** 8.03 **c)** 1.15 **d)** 0.605
3 $\mu = 40.1$ $\sigma = 11.8$
4 a) $\mu = 36$ $\sigma^2 = 30$ **b)** 0.2611

If you got them all right, skip to page 114

Normal Distribution

Improve your knowledge

The normal distribution with mean μ and variance σ^2 is written $N(\mu, \sigma^2)$ – so $N(6, 8)$ has mean 6 and variance 8 (and so standard deviation $\sqrt{8}$).

Note! The second number is the variance, not the standard deviation. You will lose a lot of marks if you get this wrong!

Questions using the normal distribution will always involve you changing between the distribution you are given in the question – which we'll call the X-values here – and the values you look up in the tables – called the z-values.

To do this, you use the formula $z = \dfrac{X - \mu}{\sigma}$, which you must learn.

The best way to avoid mistakes with the normal distribution is to always draw a diagram – just a sketch one. Do not try to save time by doing without – it is almost always a bad idea!

1 Finding Probabilities using the Normal Distribution

The key idea to remember is that probabilities correspond to areas under the curve – for example, the shaded area shown in this diagram corresponds to the probability X is less than 2.

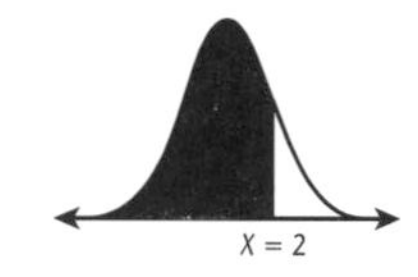

Firstly, we'll look at how to find probabilities like $X >$ something or $X <$ something.

Example 1

$X \sim N(2, 16)$. Find $P(X > 1)$

Solution

Step 1 Work out the z-value

$$z = \frac{1 - 2}{4} = -0.25$$

Note $\sigma = \sqrt{16} = 4$

Step 2 Draw a diagram showing the X- and z-values, the means of X and z and the probability you want

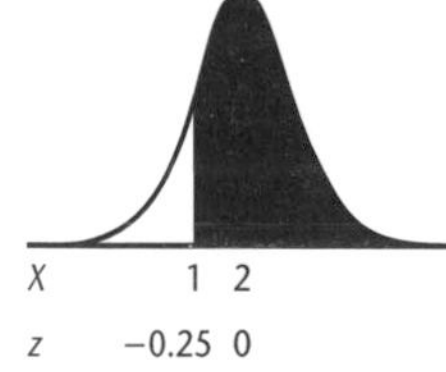

Remember to put the mean in the middle. The mean for z is always 0

Step 3 Decide whether the area you want is 'large' (more than half) or 'small' (less than half)

This is a large area

Step 4 Look up your z-value in the tables, ignoring its sign

Tables value for $z = 0.25$ is 0.5987

NEVER round tables values

Step 5 For a large area, the probability is the tables value. For a small area, it's 1 – tables value

So $P(X > 1) = 0.5987$

More Complicated Probabilities

More complicated probabilities are dealt with by finding the z-values, drawing the diagram, and then writing them as two easy probabilities added up or subtracted. You then work out the easy probabilities as shown above.

Example 2

$X \sim N(18, 20)$. Find $P(17 < X < 21)$

Solution

z-values: $\dfrac{17 - 18}{\sqrt{20}} = -0.22 \qquad \dfrac{21 - 18}{\sqrt{20}} = 0.67$

Diagram:

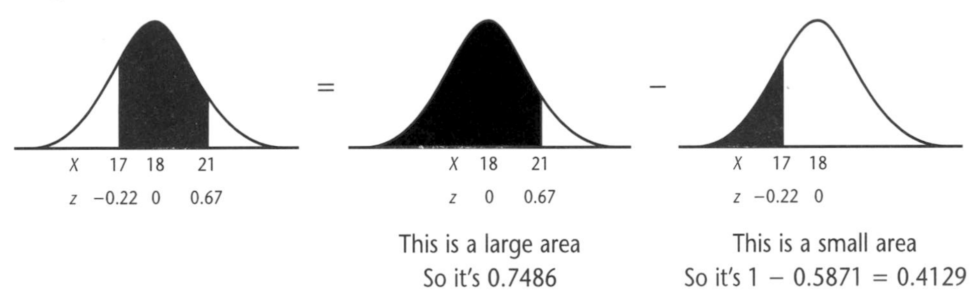

So $P(17 < X < 21) - 0.7486 - 0.4129 = 0.3357$

Finding Values in the Normal Distribution When You are Given Probabilities

The strategy here is exactly the same – draw a diagram, shade the probability you are interested in and mark on z- and X-values. The only difference is that you will know the probability and will not know an X-value.

Example 3

$X \sim N(5,25)$. $P(X < a) = 0.99$ $P(X < b) = 0.1$ Find a and b

Solution

Step 1 Draw the diagram

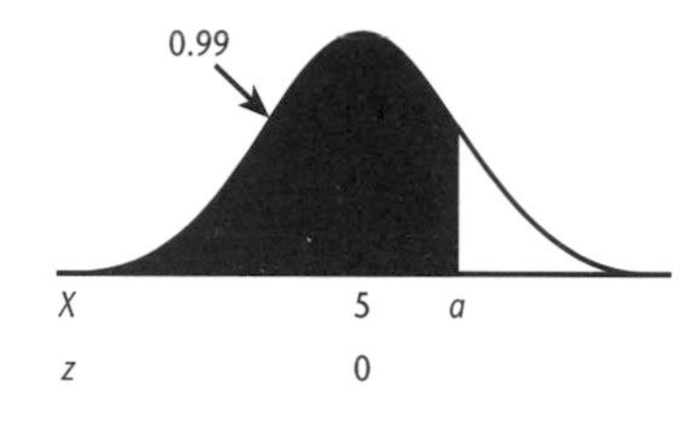

If you're not sure which side of the mean to put a, try it out both ways and check the shaded area is "sensible" – in other words, whether it's a large or small area

Step 2 Work out the large area in your diagram, if you're not already given it

We are given a large area (0.99)

Step 3 Look up the large area in the tables – remember you are using them way round! Write down the z-value you get

$z = 2.32$ is the closest

Step 4 Check from the diagram whether you have to use a positive or negative z-value

Our z-value is above zero, so we use $z = 2.32$

Step 5 Find the X-value by using the formula. Then check it is sensible from the diagram

$2.32 = \frac{a-5}{5} \Rightarrow a = 16.6$

This is sensible, as it is above 5

Now for b:

1.

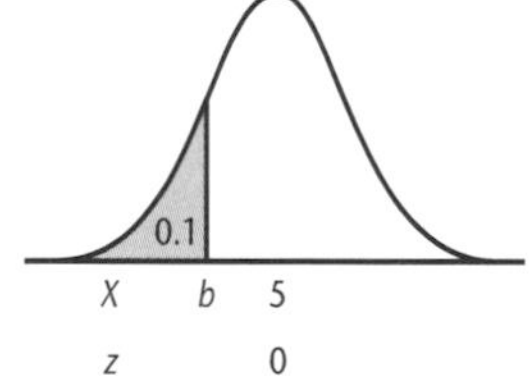

Note b must be below the mean, otherwise the shaded area would be too big

2. Large area = 1 − small area = 0.9

3. $z = 1.28$ is the closest

4. Our z-value must be below zero, so use −1.28

5. $-1.28 = \frac{b-5}{5} \Rightarrow b = -1.4$. This is sensible, as it is below 5.

large area + small area = 1 because total probability = 1

Note: Some exam boards include inverse normal tables. They give values of z corresponding to small areas – so for a small area of 0.025, they give $z = 1.96$, for example. If you have these tables, you must use them to get full marks. You can adapt the above method by always looking up the small area, rather than the big area.

Finding the Mean and Variance

Sometimes questions involving the normal distribution ask you to find the mean and/or standard deviation from the information you're given. The strategy is exactly the same as before! Note: if you are asked to find both μ and σ, you will have to solve simultaneous equations.

Example 4

The marks of students in an examination are found to be normally distributed. 2.5% of students score above 80. 15% of students score below 40. Find the mean and standard deviation of the marks.

Solution

Step 1 Put the information you have on diagrams (one for each piece)

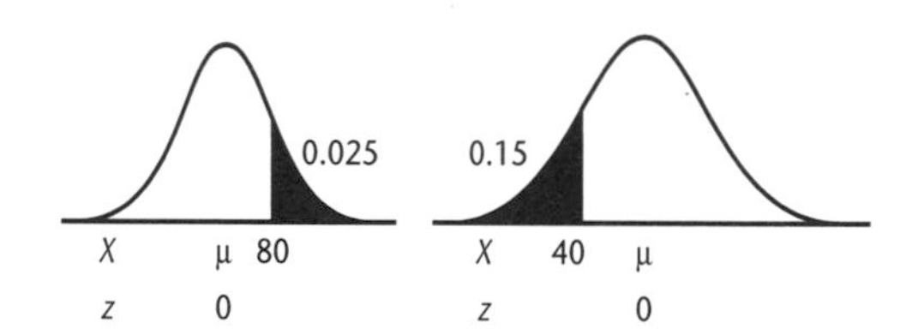

Step 2 Use the probabilities to find the z-values, as described in Section 2

We get $z = 1.96$ and 1.04

Step 3 Use the diagrams to check the signs of the z-values

$z - 1.96$ and $z = -1.04$

Step 4 Use the formula connecting z- and X-values to get equations

$1.96 = \dfrac{80 - \mu}{\sigma}$ $\quad -1.04 = \dfrac{40 - \mu}{\sigma}$

Step 5 Solve the simultaneous equations (multiply up first!)
Check your values are sensible

$1.96\sigma = 80 - \mu \quad -1.04\sigma = 40 - \mu$
Subtracting: $3\sigma = 40 \Rightarrow \sigma = 13.3$
Substitute back: $\mu = 53.9$
From the diagram, it's sensible

Using the Normal Distribution as an Approximation

You need to learn the conditions for the normal distribution to be used as an approximation, as given in the previous chapter.

40 minutes

Use your knowledge

A company produces chocolate bars. The wrapping on the chocolate bars says that their weight is 100 g. The standard deviation of the weights of the chocolate bars is known to be 5 g.

a) 99% of the chocolate bars must weigh at least as much as it says on the packet. Find the mean weight of the bars.

Draw a diagram, putting in μ and 100. Use the probability to find a z-value – it's negative!

b) Tegan buys two chocolate bars. Find the probability that exactly one of them is underweight.

Either the 1st is underweight and the 2nd isn't, or vice versa. Look in the question for probabilities!

A telephone exchange takes an average of 10 calls every minute.

Refer to the chapter on Binomial and Poisson!

a) Write down a suitable distribution to model the number of calls taken per minute, explaining why it is suitable.

Work out the original distribution first.

b) Find an approximate value for the probability that in one hour there are

Follow the method in the previous chapter – don't forget the continuity correction!

i) more than 595 calls

ii) exactly 600 calls.

Think of this as 599 < X < 601 to work out continuity correction

The lengths (in cm) of a certain type of snake follow a normal distribution with mean 30 and variance 16.

Use the method shown in section 1. Don't forget to square root the variance!

a) Find the probability a randomly selected snake is more than 35 cm long.

This is conditional probability – refer to AS work! Write down the formula for conditional probability. If it's > 35 and > 40, what must it be?

b) Given that a snake is more than 35 cm long, find the probability that it is more than 40 cm long.

Exam Practice Questions

Pure

1 $f(x) \equiv 2x^3 + x^2 - 5x + 2$

a) Factorise $f(x)$ fully.

b) Express $\dfrac{15}{f(x)}$ as partial fractions.

2 In the expansion of $(1 + x)(1 + bx)^n$, the coefficients of x and x^2 are 11 and 50 respectively. Find n and b.

3 Given that $-360° \leq x \leq 360°$, solve the equation $2\sin 2x - 3\sin x = 0$, giving your answers to the nearest degree.

4 $f(x) = 5\sin 2x + 12\cos 2x + 3$

a) Find the range of $f(x)$

b) Solve the equation $f(x) = -10 \quad -180° \leq x \leq 180°$

5 $f(x)$ is defined by: $f(x) = \dfrac{6x^2 + 18x + 10}{(3x + 2)(x + 1)^2}$; $x \in \mathbb{R}, x \neq -1, x \neq -\frac{2}{3}$

a) Express $f(x)$ in partial fractions.

b) Hence find $f'(x)$

c) Find the exact value of $\displaystyle\int_1^2 f(x)dx$

d) Express $f(x)$ as a binomial expansion in ascending powers of x, up to terms in x^2. Hence state the range of values of x for which this binomial series is valid.

e) Find the percentage error made in using the binomial expansion in part d) to estimate the value of $f(0.1)$. Give your answer to 2 significant figures.

6 The normal to the curve C, with equation $y^2 = 4x$, at the point P, where $y = 1$, meets the x-axis at the point A, as shown in the diagram.

a) Use implicit differentiation to find $\dfrac{dy}{dx}$, in terms of y, for the curve C.

b) Hence, find the equation of the normal.

The finite region R, is bounded by the part of the curve between the origin and the point P, the line PA and the x-axis.

The finite region R is rotated through 2π radians about the x-axis.

c) Calculate, in terms of π, the volume of the solid formed.

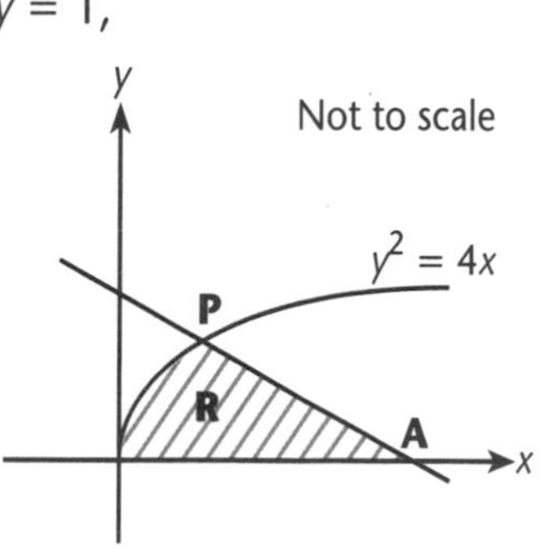

Exam Practice Questions

7 If $f(x) = e^{\sin x}$, then find:

a) $f'(x)$
b) $f''(x)$
c) the *exact values* of $\sin x$ for which $f''(x) = 0$

Mechanics

1 A batsman hits a cricket ball 0.9 metres above horizontal ground. The ball is projected at an angle of elevation $\theta°$ above the horizontal. The ball takes three seconds to move a horizontal distance of 75 metres before landing on the ground.

a) Find the vertical and horizontal components of the initial velocity of the cricket ball.
b) Hence find, to 3 significant figures, the speed and direction of the initial projection of the cricket ball.
c) Find, to 1 decimal place, the duration of time for which the cricket ball is 6 metres above horizontal ground.

2 A car of mass 800 kg can move at a maximum speed of $12\,\text{ms}^{-1}$ up a hill inclined at $\sin^{-1}(0.01)$ to the horizontal. The resistance to motion of the car is proportional to its speed, and is of magnitude 360 N when the car is moving at $12\,\text{ms}^{-1}$.

a) Find the power of the car's engine.
b) Find its maximum speed coming down the same slope.

3 In the wire framework shown below, BC = AB = 5 cm, AD = 11 cm and BE = 4 cm.

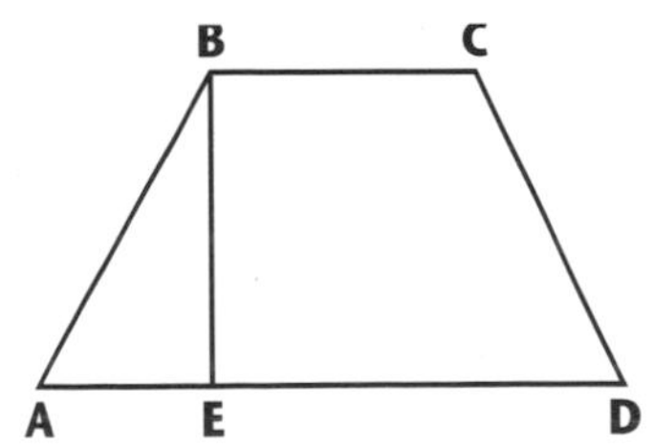

All the wire is of a uniform density and thickness.

a) Find the distance of the centre of mass of the framework from AD and from BE.

The framework is suspended from point A.

b) Find the angle that AD makes with the vertical.

Exam Practice Questions

Statistics

1 A bored student decides to throw a die 600 times. The die is biased so that the probability of an even number is three times that of an odd number.

a) Suggest a suitable distribution to model the number of even numbers the student throws.
b) Use a suitable approximation to find the probability that the student gets
   i) at least 430 even numbers
   ii) between 440 and 460 even numbers.

2 The number of breakdowns on Wayston Valley's tram network occurs at a rate of 3.5 per seven-day week. It is assumed that breakdowns occur at random, and are independent of each other.

a) Write down a suitable model with parameter(s) to model the number of breakdowns in a week.

Find the probability that there are:

b) more than 5 breakdowns in a randomly chosen week
c) no breakdowns in a randomly chosen day
d) at least 200 breakdowns in a 52-week year (use a suitable approximation).

3 At the end of training, army recruits are required to undertake a physical exercise. The times taken to complete this exercise are normally distributed with a mean of 48 minutes and a standard deviation of 5.2 minutes. Recruits who take less than 52 minutes pass the exercise. The George Marshal prize is awarded to each recruit who takes less than t minutes to complete the exercise. The George Marshall prize is awarded to only 1% of the recruits.

a) Calculate the probability that a randomly selected recruit passes the exercise.
b) Find the value of t, to 3 significant figures.

During the summer, 310 recruits complete the physical exercise. Use a suitable approximation to find the probability that:

c) fewer than 3 recruits receive the George Marshall prize
d) at least 250 recruits pass the physical exercise.

Use your Knowledge Answers

Algebra

1 a) $P = -Q + 2\ln 4$

b) $\ln y = -\ln x + 2\ln 4 \Rightarrow \ln y = \ln 4^2 - \ln x$

$\Rightarrow \ln y = \ln\left(\frac{16}{x}\right) \Rightarrow y = \frac{16}{x}$

2 a) $(1 - 4x)^7 \approx 1 - 28x + 336x^2 - 2240x^3$

b) i) $0.96^7 = (1-4x)^7 \Rightarrow 0.96 = 1 - 4x$
$\Rightarrow x = 0.01$

So $(0.96)^7 =$
$1 - 28(0.01) + 336(0.01)^2 - 2240(0.01)^3$
$= 0.75136$

ii) $0.96^7 = 0.75145$ (5DP).
So % error
$= (0.75136 - 0.75145) \div 0.75145 \times 100\%$
$= -0.012\%$

c) First find $(1 + 4x)^7 \equiv 1 + 28x + 336x^2 + 2240x^3 + \ldots$

So $(1 + 4x)^7 + (1 - 4x)^7 \approx 2 + 672x^2$

Putting $x = \frac{1}{\sqrt{3}}$: $\left(1 + 4\left(\frac{1}{\sqrt{3}}\right)\right)^7 + \left(1 - 4\left(\frac{1}{\sqrt{3}}\right)\right)^7$

$\approx 2 + 672\left(\frac{1}{\sqrt{3}}\right)^2 = 2 + 224 = 226$

3 a) Put $x = -1$: $-1 + A - B + 6 = 0$
$\Rightarrow A - B = -5$ (1)

Put $x = 2$: $8 + 4A + 2B + 6 = 60$
$\Rightarrow 2A + B = 23$ (2)

(1) + (2) $\Rightarrow 3A = 18 \Rightarrow A = 6$.
Substituting in (1): $6 - B = -5 \Rightarrow B = 11$

b) $f(x) \equiv x^3 + 6x^2 + 11x + 6$
$\equiv (x + 1)(x^2 + Ax + 6)$

$6x^2 \equiv x^2 + Ax^2 \Rightarrow A = 5$
$\Rightarrow f(x) \equiv (x + 1)(x^2 + 5x + 6)$
$\equiv (x + 1)(x + 2)(x + 3)$

4 a) $\frac{2x-5}{x-4} \equiv A + \frac{B}{x-4} \equiv \frac{A}{1} + \frac{B}{x-4}$

$\equiv \frac{A(x-4)+B}{x-4}$

$\Rightarrow 2x - 5 \equiv A(x - 4) + B$

$x = 4$: $8 - 5 = A(0) + B \Rightarrow B = 3$

$x = 0$: $-5 = -4A + B$

But $B = 3 \Rightarrow -5 = -4A + 3$

So $A = 2$

b) $\frac{2x-5}{x-4} = 2 + 3(x - 4)^{-1} = 2 + 3(-4 + x)^{-1}$

$= 2 + 3(-4)^{-1}(1 - \frac{x}{4})^{-1}$

$= 2 - \frac{3}{4}\left(1 + \frac{x}{4} + \frac{x^2}{16}\right) = \frac{5}{4} - \frac{3x}{16} - \frac{3x^2}{64}$

Valid for $-1 \le \frac{1}{4}x < 1 \Rightarrow -4 \le x < 4$

Trigonometry

1 a) LHS $\equiv \cos 4x \equiv \cos(2x + 2x)$

$\equiv \cos 2x\cos 2x - \sin 2x\sin 2x$

$\equiv \cos^2 2x - \sin^2 2x$

$\equiv (2\cos^2 x - 1)^2 - (2\sin x\cos x)^2$

$\equiv 4\cos^4 x - 4\cos^2 x + 1 - 4\sin^2 x\cos^2 x$

$\equiv 4\cos^4 x - 4\cos^2 x + 1 - 4(1 - \cos^2 x)\cos^2 x$

$\equiv 4\cos^4 x - 4\cos^2 x + 1 - 4\cos^2 x + 4\cos^4 x$

$\equiv 8\cos^4 x - 8\cos^2 x + 1 \equiv$ RHS

b) $\cos 4x = \cos^2 x \Rightarrow 8\cos^4 x - 8\cos^2 x + 1 = \cos^2 x$

$\Rightarrow 8\cos^4 x - 9\cos^2 x + 1 = 0$

$\Rightarrow (8\cos^2 x - 1)(\cos^2 x - 1) = 0$

$\Rightarrow \cos^2 x = \frac{1}{8}$ or 1

$\cos^2 x = \frac{1}{8} \Rightarrow \cos x = \pm\frac{1}{\sqrt{8}}$

$\Rightarrow x = \pm 69.3°, \pm 110.7°, \pm 249.3°, \pm 290.7°$

$\cos^2 x = 1 \Rightarrow \cos x = \pm 1$

$\Rightarrow x = 0, \pm 180°, \pm 360°$

c) Let $x = 2A$.
Then $\cos 2x = \cos^2\left(\frac{1}{2}x\right)$ $-720° \le x \le 720°$

$\Rightarrow \cos 4A = \cos^2 A$ $-360° \le x \le 360°$

So $A = \pm 69.3°, \pm 110.7°, \pm 249.3°, \pm 290.7°, 0, \pm 180°, \pm 360°$
So $x = \pm 138.6°, \pm 221.4°, \pm 498.6°, \pm 581.4°, 0, \pm 360°, \pm 720°$

2 $2\cos x\cos\frac{\pi}{3} = 2\sin x\sin\frac{\pi}{3} + 1$

$\Rightarrow 2\cos x\cos\frac{\pi}{3} - 2\sin x\sin\frac{\pi}{3} = 1$

$\Rightarrow \cos x\cos\frac{\pi}{3} - \sin x\sin\frac{\pi}{3} = \frac{1}{2}$

$\Rightarrow \cos\left(x + \frac{\pi}{3}\right) = \frac{1}{2}$

So $x + \frac{\pi}{3} = \frac{\pi}{3}, -\frac{\pi}{3} \Rightarrow x = 0, -\frac{2\pi}{3}$

3 a) i) $\tan 2A = \dfrac{2\tan A}{1 - \tan^2 A}$ so setting $A = \frac{1}{2}x$:

$$\tan x = \frac{2\tan\frac{1}{2}x}{1 - \tan^2\frac{1}{2}x} = \frac{2t}{1 - t^2}$$

ii) $\sec^2 x \equiv 1 + \tan^2 x \Rightarrow \sec^2 x \equiv 1 + \dfrac{4t^2}{(1 - t^2)^2}$

$$\equiv \frac{(1 - t^2)^2 + 4t^2}{(1 - t^2)^2} \equiv \frac{1 - 2t^2 + t^4 + 4t^2}{(1 - t^2)^2}$$

$$\equiv \frac{1 + 2t^2 + t^4}{(1 - t^2)^2} \equiv \frac{(1 + t^2)^2}{(1 - t^2)^2}$$

Hence $\sec x \equiv \sqrt{\sec^2 x} \equiv \dfrac{(1 + t^2)}{(1 - t^2)}$

b) Substituting in using t: $\dfrac{(1 + t^2)}{(1 - t^2)} = \dfrac{4t^2}{1 - t^2}$

$\Rightarrow 1 + t^2 = 4t^2 \Rightarrow 3t^2 = 1 \Rightarrow t = \pm\frac{1}{\sqrt{3}}$

$\Rightarrow \tan\left(\frac{1}{2}x\right) = \pm\frac{1}{\sqrt{3}}$

$\tan\left(\frac{1}{2}x\right) = \frac{1}{\sqrt{3}} \Rightarrow \frac{1}{2}x = 30°$;

$\tan\left(\frac{1}{2}x\right) = -\frac{1}{\sqrt{3}} \Rightarrow \frac{1}{2}x = -30°$

$\Rightarrow x = 60°, -60°$

4 $\text{LHS} \equiv \dfrac{1}{\cos x} + \dfrac{\sin x}{\cos x} \equiv \dfrac{1 + \sin x}{\cos x}$

Multiplying top and bottom by $1 - \sin x$:

$$\frac{1 + \sin x}{\cos x} \times \frac{1 - \sin x}{1 - \sin x} \equiv \frac{1 - \sin^2 x}{\cos x(1 - \sin x)}$$

$$\equiv \frac{\cos^2 x}{\cos x(1 - \sin x)} \equiv \frac{\cos x}{(1 - \sin x)} \equiv \text{RHS}$$

5 $4\sin x + 3\cos x = R\sin(x + \alpha)$
$4 = R\cos a$, $3 = R\sin\alpha$. $R = 5$; $\alpha = \tan^{-1}(0.75)$

$$f(x) = \frac{2}{6 + 5\sin(x + \alpha)}$$

$-1 \le \sin(x + \alpha) \le 1 \Rightarrow 2 \ge f(x) \ge \frac{2}{11}$

Differentiation

1 a) $u = x + 2$, $v = e^{-2x} \Rightarrow u' = 1$, $v' = -2e^{-2x}$

Product rule $\Rightarrow \dfrac{dy}{dx} = e^{-2x} - 2(x + 2)e^{-2x}$

$= e^{-2x}(1 - 2x - 4) = e^{-2x}(-2x - 3)$

b) $\dfrac{dy}{dx} = 0 \Rightarrow e^{-2x}(-2x - 3) = 0 \Rightarrow x = -\frac{3}{2}$

Hence $y = \left(-\frac{3}{2} + 2\right)e^3 = \frac{1}{2}e^3$

$\Rightarrow$ coordinates are $\left(-\frac{3}{2}, \frac{1}{2}e^3\right)$

c) $u = e^{-2x}$, $v = -2x - 3 \Rightarrow u' = -2e^{-2x}$, $v' = -2$

Product rule $\Rightarrow \dfrac{d^2y}{dx^2}$

$= e^{-2x}(4x + 6) - 2e^{-2x} = e^{-2x}(4x + 4)$
$= 4(x + 1)e^{-2x}$

At C, $\dfrac{d^2y}{dx^2} = 4\left(-\frac{1}{2}\right)e^3 = -2e^3$.

Since $\dfrac{d^2y}{dx^2} = -2e^3 < 0 \Rightarrow$ Maximum SP

2 a) $x = 2t \quad y = \dfrac{6}{t} \Rightarrow \dfrac{dx}{dt} = 2$

$\dfrac{dy}{dt} = \dfrac{-6}{t^2} \Rightarrow \dfrac{dy}{dx} = \dfrac{-3}{t^2}$

b) At $t = 3$, $x = 6$ and $y = 2$.

At $t = 3$, gradient of (T) $= \dfrac{-3}{3^2} = -\frac{1}{3}$

$\Rightarrow$ grad of Normal = 3

Equation of (N):
$y - 2 = 3(x - 6) \Rightarrow y = 3x - 16$

c) Curve = Normal $\Rightarrow$ Equate both equations

$\Rightarrow \frac{6}{t} = 3(2t) - 16$

$\frac{6}{t} = 6t - 16 \Rightarrow 6 = 6t^2 - 16t$
$\Rightarrow 6t^2 - 16t - 6 = 0 \Rightarrow 3t^2 - 8t - 3 = 0$

Factorising gives $(3t + 1)(t - 3) = 0$.

So $t = 3, -\frac{1}{3}$

We know that $t = 3$ already. So normal cuts curve again when $t = -\frac{1}{3}$

Hence $x = 2\left(-\frac{1}{3}\right) = -\frac{2}{3}$, $y = \frac{6}{t} = \frac{6}{-\frac{1}{3}} = -18$

$\Rightarrow$ The coordinates are $Q\left(-\frac{2}{3}, -18\right)$

Use your Knowledge Answers

3 a) Differentiating implicitly

$\Rightarrow 3y^2 \frac{dy}{dx} - 6 + 2y + 2x \frac{dy}{dx} = 0$

$(3y^2 + 2x) \frac{dy}{dx} = -2y + 6 \Rightarrow \frac{dy}{dx} = \frac{-2(y-3)}{(3y^2 + 2x)}$

b) $y = 2 \Rightarrow 8 - 6x + 4x - 10 = 0 \Rightarrow$
$-2x - 2 = 0 \Rightarrow x = -1 \Rightarrow$ point Q(−1, 2)

gradient of (T): $\frac{dy}{dx} = \frac{-2(2-3)}{12-2} = \frac{2}{10} = \frac{1}{5}$

$\Rightarrow$ Eqn of (T): $y - 2 = \frac{1}{5}(x + 1)$

4 $u = e^{2x}$ $v = x + 3 \Rightarrow u' = 2e^{2x}$

$v' = 1 \Rightarrow \frac{dy}{dx} = \frac{2e^{2x}(x+3) - e^{2x}}{(x+3)^2}$

Simplifying $\Rightarrow \frac{dy}{dx} = \frac{e^{2x}(2x+5)}{(x+3)^2}$.

A curve is increasing where

$\frac{dy}{dx} = \frac{e^{2x}(2x+5)}{(x+3)^2} > 0$

Since $e^{2x} > 0$ and $(x + 3)^2 > 0$, for all values of x;

then $\frac{dy}{dx} > 0 \Rightarrow 2x + 5 > 0 \Rightarrow x > -\frac{5}{2}$

Integration

1 a) $8x + 7 = (Ax + B)(4 - x) + C(2x^2 + 7)$

When $x = 4$, $39 = 39C \Rightarrow C = 1$
When $x = 0$, $7 = 4B + 7C \Rightarrow 7 - 7 = 4B$
$\Rightarrow B = 0$

When $x = 1$, $15 = 3A + 9C \Rightarrow 6 = 3A$
$\Rightarrow A = 2$

Hence $f(x) = \frac{2x}{2x^2 + 7} + \frac{2}{4 - x}$

b) $\int_1^3 \frac{2x}{2x^2+7} + \frac{1}{4-x}\, dx$

$= \left[\frac{1}{2}\ln(2x^2 + 7) - \ln(4 - x)\right]_1^3$

$= \left(\frac{1}{2}\ln 25 - \ln 1\right) - \left(\frac{1}{2}\ln 9 - \ln 3\right)$

$= \ln 5 - 0 - \ln 3 + \ln 3 = \ln 5 \Rightarrow q = 5.$

2 a) $u = x$ $\frac{dv}{dx} = \cos 3x$

$\Rightarrow \frac{du}{dx} = 1 \quad v = \frac{1}{3}\sin 3x$

$\Rightarrow \int x\cos x dx = \frac{x}{3}\sin 3x - \int \frac{1}{3}\sin 3x dx$

$= \frac{1}{3}x\sin 3x + \frac{1}{9}\cos 3x + c$

b) $x = 3\sin u \Rightarrow \frac{dx}{du} = 3\cos u \Rightarrow dx = 3\cos u du$

Change limits $\Rightarrow$ If $x = 0$, $0 = 3\sin u$
$\Rightarrow u = 0.$

When $x = \frac{3}{2}$, $\frac{3}{2} = 3\sin u$

$\Rightarrow \frac{1}{2} = \sin u \Rightarrow u = \sin^{-1}\left(\frac{1}{2}\right) = \frac{\pi}{6}$

$\int_0^{\frac{3}{2}} \sqrt{(9 - x^2)}dx = \int_0^{\frac{\pi}{6}} \sqrt{(9 - 9\sin^2 u)}\, 3\cos u du$

$= \int_0^{\frac{\pi}{6}} \sqrt{9(1 - \sin^2 u)}\, 3\cos u du$

$= \int_0^{\frac{\pi}{6}} \sqrt{9\cos^2 u}\, 3\cos u du = \int_0^{\frac{\pi}{6}} 3\cos u\, 3\cos u du$

$= 9\int_0^{\frac{\pi}{6}} \cos^2 u du \Rightarrow k = 9, a = \frac{\pi}{6},$

c) From trig, $\cos 2u = 2\cos^2 u - 1$

$\Rightarrow \cos^2 u = \frac{\cos 2u + 1}{2}.$

The integral in b) becomes:

$9\int_0^{\frac{\pi}{6}} \frac{\cos 2u + 1}{2} du \quad = \frac{9}{2}\left[\frac{\sin 2u}{2} + u\right]_0^{\frac{\pi}{6}}$

$= \frac{9}{2}\left(\left(\frac{\sqrt{3}}{4} + \frac{\pi}{6}\right) - (0)\right) = \frac{9}{8}\sqrt{3} + \frac{3}{4}\pi$

Hence $e = \frac{9}{8}$ and $f = \frac{3}{4}$.

3 a) $u = w$ $\frac{dv}{dx} = e^{-w} \Rightarrow \frac{du}{dx} = 1$

$v = -e^{-w}$

$\Rightarrow \int we^{-w}dw = -we^{-w} + \int e^{-w}dw$
$= -we^{-w} - e^{-w} + c$

b) $x = e^w \Rightarrow \frac{dx}{dw} = e^w \Rightarrow dx = e^w dw.$

Note that $x^2 = e^w e^w$

$\int \frac{3 - \ln x}{x^2} dx = \int \frac{3 - \ln e^w}{e^w e^w} e^w dw$

$= \int \frac{3 - w}{e^w} dw$, as as required.

c) $\int \frac{3 - w}{e^w} dw = \int \frac{3}{e^w} - \frac{w}{e^w} dw$

$= \int 3e^{-w} - we^{-w} dw$

Using part a), the integral becomes:

$= -3e^{-w} + we^{-w} + e^{-w} + c$

$= we^{-w} - 2e^{-w} + c$

$= \frac{w}{e^w} - \frac{2}{e^w} + c.$

Since $x = e^w$ (also $w = \ln x$), substituting back gives:

$= \frac{\ln x}{x} - \frac{2}{x} + c$

Functions

1 a) Translation $(-b, 0)$ followed by a translation $(0, a)$; or overall a translation $(-b, a)$

b)

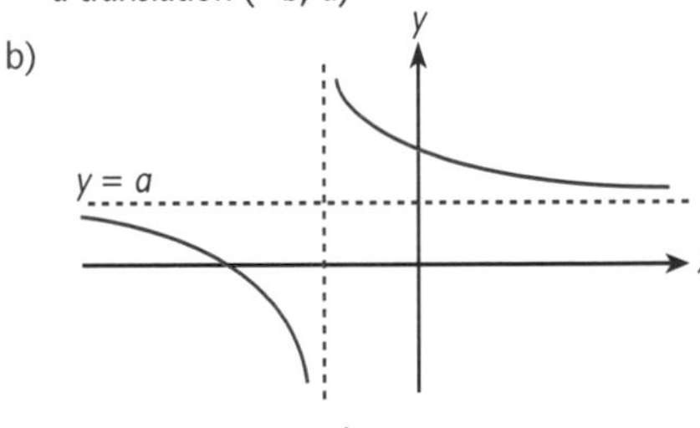

c)

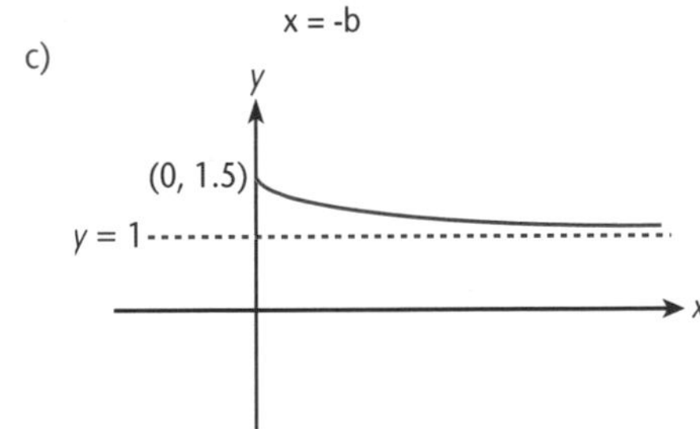

b) Asymptotes: $y = a$, $x = -b$

d) Range: $1 < f(x) \le 1.5$

e) Domain: $1 < x \le 1.5$ Range: $f^{-1}(x) \ge 0$

f) $\frac{1}{x+2} + 1 = \ln e^{2x} \Rightarrow \frac{1}{x+2} + 1 = 2x$

$\Rightarrow \frac{1}{x+2} = 2x - 1 \Rightarrow (x + 2)(2x - 1) = 1$

$\Rightarrow 2x^2 + 3x - 3 = 0$

$\Rightarrow x = \frac{-3 \pm \sqrt{9 + 24}}{4} = \Rightarrow x = \frac{-3 \pm \sqrt{33}}{4}$

$\Rightarrow x = -2.19$ (reject $\therefore < 0$), **0.69**

2 a) $gg(x) = \frac{\frac{x-1}{x+4} - 1}{\frac{x-1}{x+4} + 4} = \frac{\frac{x-1-x-4}{x+4}}{\frac{x-1+4x+16}{x+4}} = \frac{-5}{5x+15} = \frac{-1}{x+3}$

$\Rightarrow a = -1, b = 3$

b) $y = \frac{x-1}{x+4} \Rightarrow x = \frac{y-1}{y+4}$

$x(y + 4) = y - 1. \Rightarrow xy + 4x = y - 1$

$\Rightarrow xy - y = -1 - 4x$

$\Rightarrow y(x - 1) = -1 - 4x$

$\Rightarrow y = \frac{-1-4x}{x-1} = \frac{4x+1}{1-x}$

Hence $g^{-1}(x) = \frac{4x+1}{1-x}$, $x \in \mathbb{R}, x \neq 1$

Vectors

1 a) $\mathbf{v}.(\mathbf{i} - 2\mathbf{k}) = 0 \Rightarrow a - 2c = 0 \Rightarrow a = 2c$

$\mathbf{v}.(2\mathbf{i} + \mathbf{j} - \mathbf{k}) = 0 \Rightarrow 2a + b - c = 0$

$\Rightarrow 4c + b - c = 0 \Rightarrow b = -3c$

b) $\mathbf{v} = 2c\mathbf{i} - 3c\mathbf{j} + c\mathbf{k}$

$\Rightarrow |\mathbf{v}| = \sqrt{(4c^2 + 9c^2 + c^2)}$

$= \sqrt{56}$

$c\sqrt{14} = \sqrt{56} \Rightarrow c\sqrt{14} = \sqrt{4}\sqrt{14} \Rightarrow c = 2$

2 a) $\mathbf{r} = \mu(-5\mathbf{i} + \mathbf{k})$

b) Equation of L is $\mathbf{r} = 2\mathbf{i} + 4\mathbf{j} - \mathbf{k} + \lambda(\mathbf{i} - 2\mathbf{j} + 3\mathbf{k})$

$(-5\mathbf{i} + \mathbf{k}).(\mathbf{i} - 2\mathbf{j} + 3\mathbf{k})$

$= \sqrt{[(-5)^2 + 1^2]}\sqrt{[1^2 + (-2)^2 + 3^2]} \cos\theta$

$-5 + 3 = \sqrt{26}\sqrt{14} \cos\theta \Rightarrow \theta = 96.0°$

But this is angle between lines, so we need acute angle $\Rightarrow 180° - 96° = 84°$

3 a) i) $\overrightarrow{OQ} = (5 + \mu)\mathbf{i} + (2 - \mu)\mathbf{j} - \mu\mathbf{k}$

We need this to be perpendicular to the line. Direction of line is given by its direction vector – so we need $\overrightarrow{OQ}$ perpendicular to direction vector of line.

So $((5+\mu)\mathbf{i} + (2-\mu)\mathbf{j} - \mu\mathbf{k}).(\mathbf{i} - \mathbf{j} - \mathbf{k}) = 0$

$5 + \mu - 2 + \mu + \mu = 0 \Rightarrow \mu = -1$

So $\overrightarrow{OQ} = 4\mathbf{i} + 3\mathbf{j} + \mathbf{k}$

ii) As the sketch shows, this shortest distance is OQ

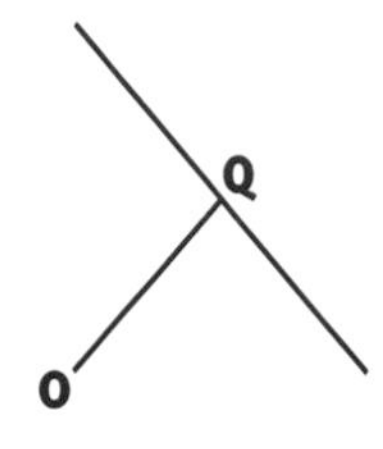

$|\overrightarrow{OQ}| = \sqrt{(4^2 + 3^2 + 1^2)} = \sqrt{26}$

b) $\overrightarrow{OQ} = \overrightarrow{QR} \Rightarrow \overrightarrow{OR} = 2\overrightarrow{OQ} = 8\mathbf{i} + 6\mathbf{j} + 2\mathbf{k}$

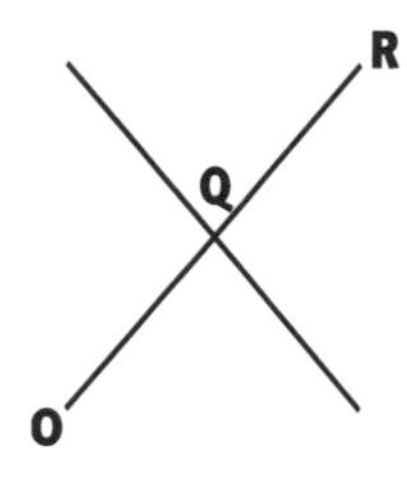

Projectiles

1 a)

A
12 ms^{-1}
80 m
B

AB (↓):
$u = 12 \sin 15°$, $s = 80$, $a = 9.8$, $t = ??$, $v = ??$

$s = ut + \frac{1}{2}at^2$

$\Rightarrow 80 = (12 \sin 15)t + \frac{1}{2}(9.8)t^2$

Hence: $4.9t^2 + (12 \sin 15)t - 80 = 0$

$\Rightarrow t = \frac{-12\sin 15 \pm \sqrt{(12\sin 15)^2 - (4 \times 4.9 \times -80)}}{2 \times 4.9}$

$\Rightarrow t = \frac{-12\sin 15 \pm 39.71959...}{9.8}$

$= -4.370..., 3.736...$. Hence $t = 3.7$ s (2sf)

b) AB (→): $s = ??$, $u = 12 \cos 15°$, $t = 3.736....$
$\therefore s = ut \Rightarrow s = (12 \cos 15) \times 3.736...$

$s = 43.3... \Rightarrow s = 43$ m (2sf)

2 a) AB (↑): $u = ??$, $s = 3.97$, $a = -9.8$, $t = ??$, $v = 0$. Hence $v^2 = u^2 + 2as$

$0 = u^2 + 2(-9.8)(3.97) \Rightarrow u^2 = 77.812$
$\Rightarrow u_V = \sqrt{77.812} = 8.8211...$ ms^{-1}.

Also, $v = u + at \Rightarrow 0 = 8.8211... + (-9.8)t$

$\Rightarrow t = \frac{8.8211...}{9.8} = 0.90011.. = 0.900$ s (3sf)

b) AB(→): $s = 21.2$, $u_H = ??$, $t = 0.90011...$

$\Rightarrow s = ut \Rightarrow u_H = \frac{21.2}{0.90011...} = 23.553...$ ms^{-1}

AC (→): $s = 5.9$, $u_H = 25.553...$, $t = ??$

$\Rightarrow s = ut \Rightarrow t = \frac{s}{u} = \frac{5.9}{23.533...} = 0.25050...$ s

AC (↑): $u_V = 8.8211...$, $s = ??$, $a = -9.8$,
$t = 0.25050...$, $v = ??$

Hence: $s = ut + \frac{1}{2}at^2$

$\Rightarrow s = (8.8211)(0.25050) + \frac{1}{2}(-9.8)(0.25050)^2$
$= 1.9022...$ m

Since 1.90 m > 1.8 m, the particle passes over the man's head.

3

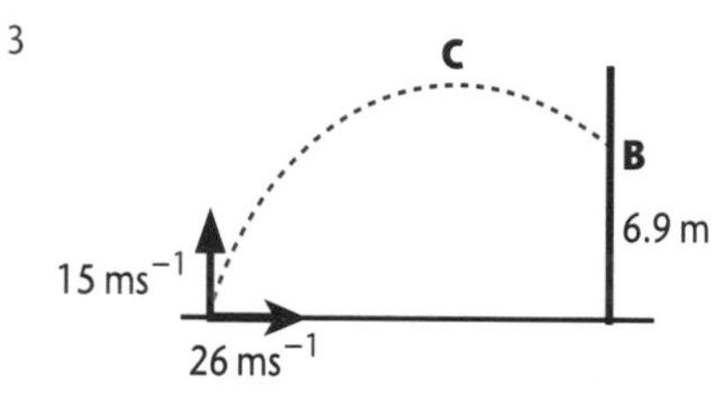

Let C be the position of the particle's maximum height.

a) AC (↑): $u = 15$, $s = ??$, $a = -9.8$, $t = ??$, $v = 0$

Hence: $v^2 = u^2 + 2as \Rightarrow 0 = 225 + 2(-9.8)s$

$19.6s = 225 \Rightarrow s = \frac{225}{19.6} = 11.480... = 11.5$m (3sf)

b) AB (↑): $u = 15$, $s = 6.9$, $a = -9.8$, $t = ??$, $v = ??$
Hence: $s = ut + \frac{1}{2}at^2$

$\Rightarrow 6.9 = 15t + \frac{1}{2}(-9.8)t^2$

$\Rightarrow 4.9t^2 - 15t + 6.9 = 0$

Formula: $t = \frac{15 \pm \sqrt{225 - (4 \times 4.9 \times 6.9)}}{2 \times 4.9}$

$= \frac{15 \pm \sqrt{89.76}}{9.8} = 0.564..., 2.497...$

Since particle hits wall after maximum height, we take the larger time. Hence $t = 2.50$ s (3sf)

c) At B, $v_H = 26\,ms^{-1}$ (remains the same).
AB (↑): $\Rightarrow v_V = u + at$
$= 15 + (-9.8)(2.497...)$
$v_V = -9.4706...\ ms^{-1}$.
Hence speed $= \sqrt{26^2 + (-9.4706...)^2}$
$= 27.67... = 27.7\,ms^{-1}$ (3sf)

Work, energy and Power

1 a) i) Car: $D - R_{car} - T = 0$
Trailer $T - R_{trailer} = 0$

Combining: $D = R_{car} + R_{trailer}$
Also: $D = 10000 \div 20 = 500\,N$
So $R_{car} + R_{trailer} = 500$

Resistance proportional to mass, so

$$R_{car} = \frac{1200}{1200 + 800} \times 500 = 300\,N$$

ii) $500 - 300 - T = 0 \Rightarrow T = 200\,N$

b) Car: $D - T - 300 + 1200g\sin5° = 1200a$
Trailer $T - 200 + 800g\sin5° = 800a$
Combining: $D - 500 + 2000g\sin5° = 2000a$

Also: $D = 10000 \div 20 = 500\,N$
So $a = g\sin5° = 0.854\,ms^{-2}$

2 Initial energy $= \frac{1}{2} \times 0.06 \times 4^2$

Final energy $= 0.06 \times 9.8 \times d\sin30°$

Frictional force $= \mu R = 0.1 \times 0.06 \times 9.8 \times \cos30°$

Work done against friction $= 0.0588\cos30° \times d$

So $0.0588\cos30° \times d = \frac{1}{2} \times 0.06 \times 4^2 - 0.06 \times 9.8 \times d\sin30°$

$d(0.0588\cos30° + 0.588\sin30°) = 0.48$

$d = 1.39\,m$

Centres of Mass

1 a)

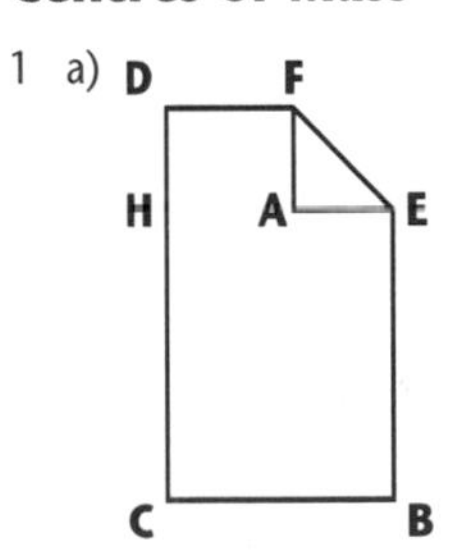

$$25 \times \frac{20}{3} + 25 \times 2.5 + 150 \times 5 = 200 \times a$$
$$\Rightarrow a = 4.896$$

Body	Mass	Distance of centre of mass from CD	Distance of centre of mass from BC
AEF	25	$5 + \frac{5}{3}$	$15 + \frac{5}{3}$
AFDH	25	2.5	17.5
BCHE	150	5	7.5
whole thing	200	a	b

$$25 \times \frac{50}{3} + 25 \times 17.5 + 150 \times 7.5 = 200 \times b$$
$$\Rightarrow b = 9.896$$

So centre of mass is 4.896 cm from CD, 9.896 cm from BC

b)

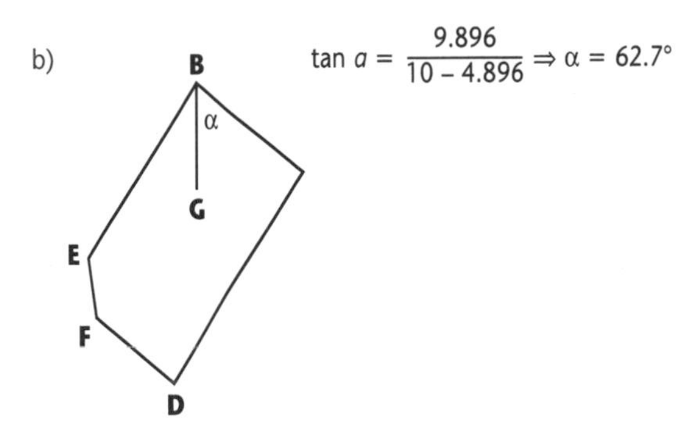

$$\tan a = \frac{9.896}{10 - 4.896} \Rightarrow \alpha = 62.7°$$

2 a)

Body	Mass	Distance of centre of mass from AB	Distance of centre of mass from Line through A perpendicular to AB
Small semicircle	$0.5\pi R^2$	$\frac{4R}{3\pi}$	R
Lamina	$1.5\pi R^2$	a	b
Big semicircle	$2\pi R^2$	$\frac{8R}{3\pi}$	2R

$$0.5\pi R^2 \times \frac{4R}{3\pi} + 1.5\pi R^2 \times a = 2\pi R^2 \times \frac{8R}{3\pi} \Rightarrow a = \frac{28R}{9\pi}$$

$$0.5\pi R^2 \times R + 1.5\pi R^2 \times b = 2\pi R^2 \times 2R \Rightarrow b = \frac{7R}{3}$$

So centre of mass is $\frac{28R}{9\pi}$ from AB,

and $\frac{7R}{3}$ from line through A perpendicular to AB

Use your Knowledge Answers

b)

Body	Mass	Distance of centre of mass from	
		AB	Line through A perpendicular to AB
Lamina	M	$\frac{28R}{9\pi}$	$\frac{7R}{3}$
Particle	$2M$	0	$4R$
Whole thing	$3M$	c	d

$$M \times \frac{28R}{9\pi} + 2M \times 0 = 3M \times c \Rightarrow c = \frac{28R}{27\pi}$$

$$M \times \frac{7R}{3} + 2M \times 4R = 3M \times d$$

$$\Rightarrow d = \frac{31R}{9}$$

So centre of mass is $\frac{28R}{27\pi}$ from AB,

and $\frac{31R}{9}$ from line through A perpendicular to AB

Moments

1 a)

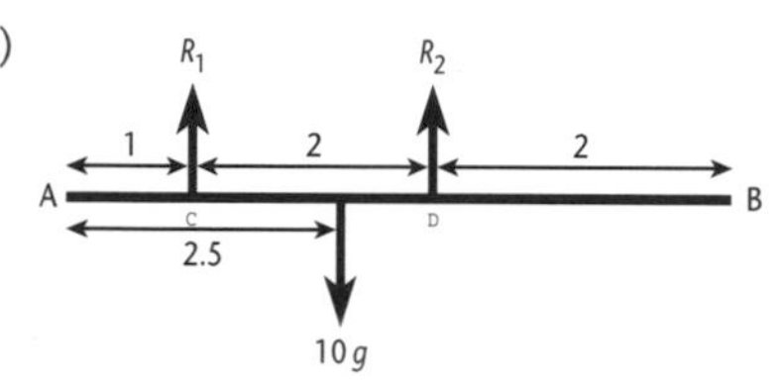

Moments about C:
$1.5 \times 10g = 2 \times R_2 \Rightarrow R_2 = 7.5g$
Resolving vertically: $R_1 + R_2 = 10g \Rightarrow R_1 = 2.5g$

b) Plank starts to tilt ⇒ loses contact at C ⇒ $R_1 = 0$. So we have:

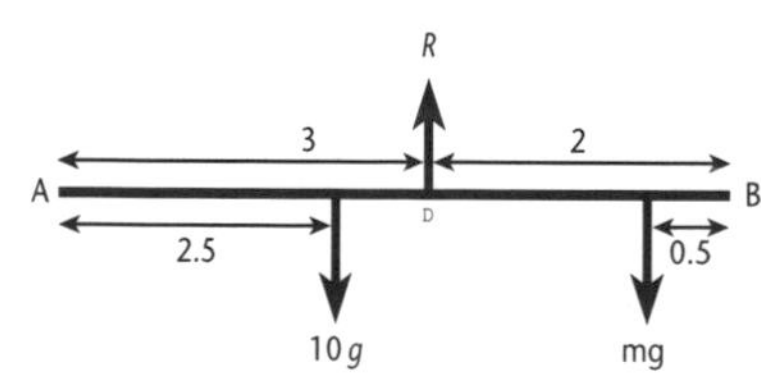

$$0.5 \times 10g = 1.5 \times mg \Rightarrow m = \tfrac{10}{3}\text{ kg}$$

c) Plank is modelled as a rod
Cat is modelled as a particle

2 a)

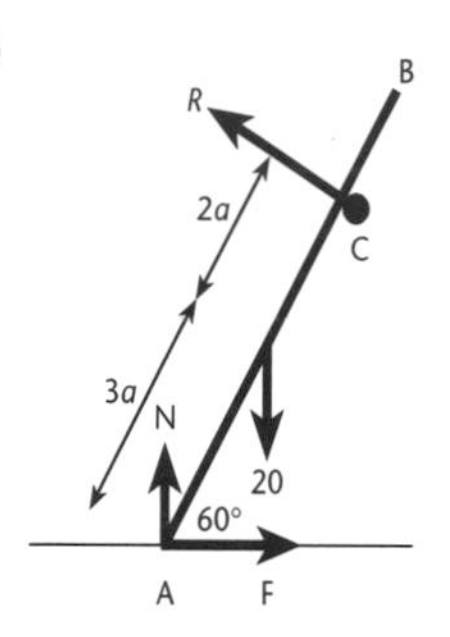

Taking moments about A:
$20\sin 30° \times 3a = R \times 5a$

$$R = 6\text{N}$$

b) Resolving vertically: $20 = N + R\cos 60°$
$\Rightarrow N = 17\text{N}$
horizontally: $F = R\sin 60° \Rightarrow F = 3\sqrt{3}\text{N}$

Friction limiting $\Rightarrow F = \mu R \Rightarrow \mu = \frac{3\sqrt{3}}{17}$

Particle Kinematics

1 a) $v = \frac{ds}{dt} = 3t^2 - 14t + 8 = 0$ (at rest)

$\Rightarrow (3t - 2)(t - 4) = 0 \Rightarrow t = \tfrac{2}{3}\text{s}, 4\text{s}$

b) When $t = 0$s, $s = 4$m. When $t = \tfrac{2}{3}$s,

$s = \tfrac{8}{27} - \tfrac{28}{9} + \tfrac{16}{3} + 4 = \tfrac{176}{27} = 6\tfrac{14}{27}\text{m}$

When $t = 4$s,

$s = 64 - 112 + 32 + 4 = -12$m.

When $t = 6$s, $s = 216 - 252 + 48 + 4 = 16$m

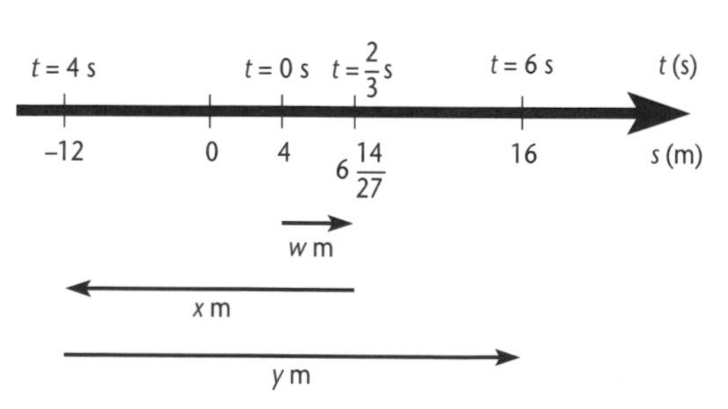

From the diagram, $w = 2\tfrac{14}{27}$m, $x = 18\tfrac{14}{27}$m, $y = 28$m

∴ Distance covered $= w + x + y = 49\tfrac{1}{27}$m

= 49 m (nearest metre)

Use your Knowledge Answers

2 a) $\mathbf{v} = \int \mathbf{a}\,dt = 3t\mathbf{i} + (t^2 - t)\mathbf{j} + \mathbf{c}_1$. BCs: $t = 1$ s,
$\mathbf{v} = 27\mathbf{i} + 6\mathbf{j} \Rightarrow 27\mathbf{i} + 6\mathbf{j} = 3\mathbf{i} + 0\mathbf{j} + \mathbf{c}_1$
$\Rightarrow \mathbf{c}_1 = 24\mathbf{i} + 6\mathbf{j}$.
Hence, $\mathbf{v} = (3t + 24)\mathbf{i} + (t^2 - t + 6)\mathbf{j}\,\text{ms}^{-1}$

b) i) parallel to $2\mathbf{i} + \mathbf{j} \Rightarrow$ ratio of $\mathbf{i}$ to $\mathbf{j}$ component of velocity is 2:1 $\therefore$

$$\frac{3t + 24}{t^2 - t + 6} = \frac{2}{1}$$

$3t + 24 = 2t^2 - 2t + 12 \Rightarrow 2t^2 - 5t - 12$
$= 0 \Rightarrow (2t + 3)(t - 4) = 0$
$\Rightarrow t = -\frac{3}{2}, 4 \Rightarrow t = 4\,\text{s}$

c) $\mathbf{r} = \int \mathbf{v}\,dt = \left(\frac{3t^2}{2} + 24t\right)\mathbf{i} + \left(\frac{t^2}{3} - \frac{t^2}{2} + 6t\right)\mathbf{j} + \mathbf{c}_2$

BCs: $\mathbf{r} = 0,\ t = 0 \Rightarrow 0 = 0\mathbf{i} + 0\mathbf{j} + \mathbf{c}_2$
$\Rightarrow \mathbf{c}_2 = 0\mathbf{i} + 0\mathbf{j}$.

So, $\mathbf{r} = \left(\frac{3t^2}{2} + 24t\right)\mathbf{i} + \left(\frac{t^2}{3} - \frac{t^2}{2} + 6t\right)\mathbf{j}$

When $t = 1$ s;

$\mathbf{r} = (\frac{3}{2} + 24)\mathbf{i} + (\frac{1}{3} - \frac{1}{2} + 6)\mathbf{j} = \frac{51}{2}\mathbf{i} + \frac{35}{6}\mathbf{j}$

Distance OP $= |\mathbf{r}| = \sqrt{\left(\frac{51}{2}\right)^2 + \left(\frac{35}{6}\right)^2}$
$= 26.159\ldots = 26.2\text{m (3sf)}$

3 a) $\mathbf{F} = m\mathbf{a} \Rightarrow (8t - 3)\mathbf{i} + 2t\mathbf{j} = 4\mathbf{a}$

$\Rightarrow \mathbf{a} = \left(2t - \frac{3}{4}\right)\mathbf{i} + \frac{t}{2}\mathbf{j}$

$\mathbf{v} = \int \mathbf{a}\,dt = \left(t^2 - \frac{3}{4}t\right)\mathbf{i} + \frac{t^2}{4}\mathbf{j} + \mathbf{c}_1$

BCs: $t = 0,\ \mathbf{v} = -\frac{5}{2}\mathbf{i} - \mathbf{j}$

$\Rightarrow -\frac{5}{2}\mathbf{i} - \mathbf{j} = 0\mathbf{i} + 0\mathbf{j} + \mathbf{c}_1$

$\Rightarrow \mathbf{c}_1 = -\frac{5}{2}\mathbf{i} - \mathbf{j}$

Hence: $\mathbf{v} = \left(t^2 - \frac{3}{4}t - \frac{5}{2}\right)\mathbf{i} + \left(\frac{t^2}{4} - 1\right)\mathbf{j}\ \text{ms}^{-1}$

b) $\mathbf{i} : = 0 \Rightarrow t^2 - \frac{3}{4}t - \frac{5}{2} = 0 (\times 4)$
$\Rightarrow 4t^2 - 3t - 10 = 0 \Rightarrow (4t + 5)(t - 2)$
$= 0 \Rightarrow t = -\frac{5}{4}, 2$

$\therefore t = 2\text{s}$. $\mathbf{j}$: $= 0 \Rightarrow \frac{t^2}{4} =- 1 = 0\ (\times 4)$

$\Rightarrow t^2 - 4 = 0 \Rightarrow (t + 2)(t - 2)$
$= 0 \Rightarrow t = -2, 2 \therefore t = 2\text{s}$

Since both the $\mathbf{i}$ and $\mathbf{j}$ components of $\mathbf{v}$ are 0, the particle is instantaneously at rest, when $t = 2$s

c) $\mathbf{r} = \int \mathbf{v}\,dt = \left(\frac{t^3}{3} - \frac{3}{8}t^2 - \frac{5}{2}t\right)\mathbf{i} + \left(\frac{t^3}{12} - t\right)\mathbf{j} + \mathbf{c}_2$.

BCs: $t = 0,\ \mathbf{r} = 0 \Rightarrow 0 = 0\mathbf{i} + 0\mathbf{j} + \mathbf{c}_2 \Rightarrow \mathbf{c}_2 = 0$

Hence: $\mathbf{r} = \left(\frac{t^3}{3} - \frac{3}{8}t^2 - \frac{5}{2}t\right)\mathbf{i} + \left(\frac{t^3}{12} - t\right)\mathbf{j}$

When $t = 1$ s,

$\mathbf{r} = \left(\frac{1}{3} - \frac{3}{8} - \frac{5}{2}\right)\mathbf{i} + \left(\frac{1}{12} - 1\right)\mathbf{j} = -\frac{61}{24}\mathbf{i} - \frac{11}{12}\mathbf{j}\ \therefore A(-\frac{61}{24}, -\frac{11}{12})$

When $t = 2$s, $\mathbf{r} = \left(\frac{8}{3} - \frac{3}{2} - 5\right)\mathbf{i} + \left(\frac{8}{12} - 2\right)\mathbf{j} \Rightarrow B\left(-\frac{23}{6}, -\frac{4}{3}\right)$

Hence: distance AB

$= \sqrt{\left(-\frac{61}{24} - -\frac{23}{6}\right)^2 + \left(-\frac{11}{12} - -\frac{4}{3}\right)^2}$

$= 1.3572\ldots = 1.36\text{m (3sf)}$

Binomial and Poisson Distributions

1 X = number of telephone calls

a) $X \sim \text{Poi}(4.75)$ for 15 mins $\Rightarrow P(X = 6)$

$= \frac{e^{-4.75}(4.75)^6}{6!} = 0.138\text{ (3sf)}$

b) $X \sim \text{Poi}(9.5)$ for 30 mins
$\Rightarrow P(X \geq 10) = 1 - P(X \leq 9) = 1 - 0.5218$
$= 0.4782 = 0.478\text{ (3sf)}$

c) $X \sim \text{Poi}(57)$ for 3 hours $\rightarrow X \sim N(57, 57)$
Hence $P(X > 50) \cong P(X > 50.5)$ (by c.c.)

z-value: $\frac{50.5 - 57}{\sqrt{57}} = -0.861\ldots$
$\Rightarrow P(Z > -0.86) = \Phi(0.86)$
$= 0.8051 = 0.805$ (3sf) (0.805)

d) prob $= (0.1380)^2 = 0.01904\ldots = 0.0190$ (3sf)

2 i) a) W = number who pay by debit card
$\Rightarrow W \sim \text{Bin}(25, 0.40)$
$P(W > 10) = 1 - P(W \leq 10) = 1 - 0.5858$
$= 0.4142 = 0.414$ (3sf)

b) X = number who pay by cash
$\Rightarrow X \sim \text{Bin}(25, 0.55)$

Either: $P(X < 14) = P(X \leq 13) = 0.4574$
$= 0.457$ (3sf) (for some boards)

Or: Y = number who do not pay by cash
$\Rightarrow Y \sim \text{Bin}(25, 0.45)$

$P(X < 14) = P(Y \geq 12) = 1 - P(Y \leq 11)$
$= 1 - 0.5426 = 0.4574 = 0.457$ (3sf) (for other boards)

Use your Knowledge Answers

c) From part b) $P(Y > 12) = 1 - P(Y \le 12)$
$= 1 - 0.6937 = 0.3063 = 0.306$ (3sf)

d) $E(W) = np = 25 \times 0.40 = 10$

ii) a) C = number who pay by cheque
$\Rightarrow C \sim \text{Bin}(80, 0.05) \rightarrow C \sim \text{Poi}(4)$

$P(C < 5) = P(C \le 4) = 0.6288 = 0.629$ (3sf)

b) $X \sim \text{Bin}(80, 0.55) \rightarrow X \sim N(44, 19.8)$ Hence

$P(48 \le X \le 53) \cong P(47.5 < X < 53.5)$ (by c.c.)

z-values: $\dfrac{47.5 - 44}{\sqrt{19.8}} = 0.787...$

$\dfrac{53.5 - 44}{\sqrt{19.8}} = 2.135...$

$\Rightarrow P(0.79 < Z < 2.13)$
$= \Phi(2.13) - \Phi(0.79)$
$= 0.9834 - 0.7852$
$= 0.1982 = 0.198$ (3sf)

3 i) X = number of birthday cakes sold
$\Rightarrow X \sim \text{Poi}(6)$ for 1 day

a) Formula $\Rightarrow P(X = 4) = \dfrac{e^{-6}(6)^4}{4!}$

$= 0.1339... = 0.134$ (3sf)

or tables $\Rightarrow P(X = 4) = P(X \le 4) - P(X \le 3)$
$= 0.2851 - 0.1512 = 0.1339$
$= 0.134$ (3sf)

b) $P(X \ge 6) = 1 - P(X \le 5) = 1 - 0.4457$

$= 0.5543 = 0.554$ (3sf)

ii) Let Y = number of days when at least 6 birthday cakes are sold $\Rightarrow Y \sim \text{Bin}(5, 0.5543)$

$P(Y = 3) = {}^5C_3(0.5543)^3(0.4457)^2 = 0.33831...$
$= 0.338$ (3sf)

Normal Distribution

1 a) $-2.32 = \dfrac{100 - \mu}{5} \Rightarrow \mu = 111.6\text{g}$

b) $2 \times 0.99 \times 0.01 = 0.0198$

2 a) Poi(10). It is reasonable to assume calls arrive at a uniform rate, independently and singly

b) Use N(600, 600)

i) P(>595 calls) $\Rightarrow P(X > 595.5)$.

$z = \dfrac{595.5 - 600}{\sqrt{600}} = -0.18$

So $P(X > 595.5) = 0.5714$

ii) P(599 < calls < 601)
$\Rightarrow P(599.5 < X < 600.5)$

z-values: $\dfrac{599.5 - 600}{\sqrt{600}} = -0.02$

$\dfrac{600.5 - 600}{\sqrt{600}} = 0.02$

So probability $= 0.5080 - (1 - 0.5080) = 0.0160$

3 a) $z = \dfrac{35 - 30}{4} = 1.25$

$\Rightarrow$ probability $= 1 - 0.8944 = 0.1056$

b) $P(X > 40 \mid X > 35) = \dfrac{P(X > 40 \text{ and } X > 35)}{P(X > 35)}$

$= \dfrac{P(X > 40)}{P(X > 35)}$

To find $P(X > 40)$:

$z = \dfrac{40 - 30}{4} = 2.5$

$\Rightarrow$ probability $= 1 - 0.9938 = 0.0062$

So $P(X > 40 \mid X > 35) = \dfrac{0.0062}{0.1056} = 0.0587$

Pure

1 a) $(x - 1)(2x - 1)(x + 2)$ b) $\dfrac{5}{x-1} - \dfrac{12}{2x-1} + \dfrac{1}{x-2}$

2 $n = 5$, $b = 2$

3 $0°, \pm180°, \pm360°, \pm41°, \pm319°$

4 a) $-10 \le f(x) \le 16$ b) $101°, -79°$,

5 a) $f(x) = \dfrac{6x^2 + 18x + 10}{(3x + 2)(x + 1)^2} = \dfrac{A}{(3x - 2)} + \dfrac{B}{(x + 1)} + \dfrac{C}{(x - 1)^2}$

Hence: $6x^2 + 18x + 10$
$= A(x + 1)^2 + B(3x + 2)(x + 1) + C(3x + 2)$

Let $x = -1 \Rightarrow -2 = -C \Rightarrow C = 2$

Let $x = -\frac{2}{3} \Rightarrow \frac{8}{3} - 12 + 10 = \frac{A}{9} \Rightarrow \frac{2}{3} = \frac{A}{9} \Rightarrow A = 6$

Let $x = 0 \Rightarrow 10 = A + 2B + 2C \Rightarrow 10 = 6 + 2B + 4$
$\Rightarrow B = 0.$

Hence, $f(x) = \dfrac{6}{(3x + 2)} + \dfrac{2}{(x + 1)^2}$
$= 6(3x + 2)^{-1} + 2(x + 1)^{-2}$

b) $f'(x) = -6(3x + 2)^{-2}\,3 + (-4)(x + 1)^{-3}\,1$

$= \dfrac{-18}{(3x + 2)^2} - \dfrac{4}{(x + 1)^3}$

c) $\int_1^2 f(x)dx = 2\int_1^2 \frac{3}{3x+2}\,dx + 2\int_1^2 (x+1)^{-2}dx$

$= \left[2\ln(3x+2) - \frac{2}{x+1}\right]_1^2$

$= \left(2\ln 8 - \frac{2}{3}\right) - (2\ln 5 - 1) = 2\ln\left(\frac{8}{5}\right) + \frac{1}{3}$

d) $f(x) = 6(2+3x)^{-1} + 2(1+x)^{-2}$

$= 3\left(1 + \frac{3}{2}x\right)^{-1} + 2(1+x)^{-2}$

$= 3\left[1 + (-1)\left(\frac{3x}{2}\right) + \frac{(-1)(-2)}{2!}\left(\frac{3x}{2}\right)^2 + \ldots\right]$

$+ 2\left[1 + (-2)(x) + \frac{(-2)(-3)}{2!}(x)^2 + \ldots\right]$

$= 3\left[1 - \frac{3x}{2} + \frac{9x^2}{4}\right] + 2[1 - 2x + 3x^2]$

$= 5 - \frac{17}{2}x + \frac{51}{4}x^2$ $(= 5 - 8.5x + 12.75x^2 + \ldots)$

$(2+3x)^{-1}$ is valid when $\left|\frac{3x}{2}\right| < 1 : |x| < \frac{2}{3}$.

$(1+x)^{-2}$ is valid when $|x| < 1$

Hence binomial expansion is valid when both inequalities coincide, i.e. when, $|x| < \frac{2}{3}$ or $-\frac{2}{3} < x < \frac{2}{3}$.

e) Estimated: $f(0.1) \approx 5 - 8.5(0.1) + 12.75(0.1)^2$

$= 5 - 0.85 + 0.1275 = 4.2775$

Calculated: $f(0.1) = \frac{0.06 + 1.8 + 10}{(2.3)(1.1)^2} = \frac{11.86}{2.783}$

$= 4.261588214\ldots$

%age error $= \frac{error}{actual} \times 100$

$= \frac{|4.2775 - 4.261588214\ldots|}{4.261588214\ldots} \times 100$

$= 0.373\ldots$

$= 0.37\%$ (2sf)

6 a) $y^2 = 4x \Rightarrow 2y \cdot \frac{dy}{dx} = 4 \Rightarrow \frac{dy}{dx} = \frac{4}{2y} \Rightarrow = \frac{2}{y}$

b) $y = 1 \Rightarrow \frac{dy}{dx} = 2 \Rightarrow$ gradient of Normal $= -\frac{1}{2}$.

$y = 1 \Rightarrow 1 = 4x \Rightarrow x = \frac{1}{4} \Rightarrow P\left(\frac{1}{4}, 1\right)$

Eqn of normal: $y - 1 = -\frac{1}{2}\left(x - \frac{1}{4}\right) \Rightarrow y - 1 = -\frac{1}{2}x + \frac{1}{8}$

$\Rightarrow y = -\frac{1}{2}x + \frac{9}{8}$

c)

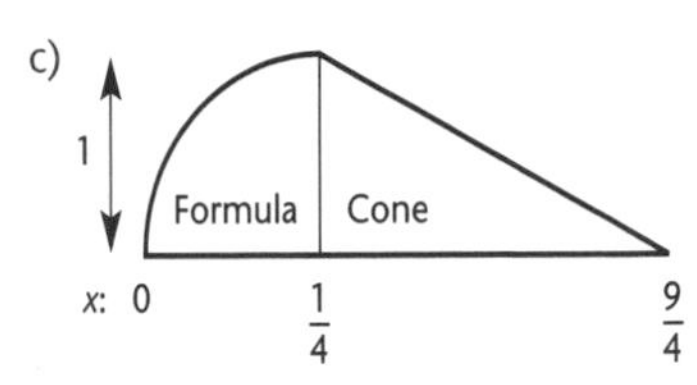

The region R is split into two. The first region is a region enclosed between axis and curve $\Rightarrow$ Use formula

$\therefore$ Vol $= \pi \int_0^{0.25} y^2\,dx = \pi \int_0^{0.25} 4x\,dx = \pi[2x^2]_0^{0.25}$

$= 2\pi\left[\frac{1}{16} - 0\right] = \frac{\pi}{8}$(units)^3.

The second region is a triangle, which rotates to a cone.

So Vol. cone $= \frac{1}{3}\pi r^2 h = \frac{1}{3}\pi\, 1^2\left(\frac{9}{4} - \frac{1}{4}\right) = \frac{2}{3}\pi$ (units)^3.

Total $V = \frac{\pi}{8} + \frac{2\pi}{8} = \frac{19}{24}\pi$ (units)^3.

7 a) $f'(x) = \cos x\, e^{\sin x}$

b) $u = \cos x,\ v = e^{\sin x} \Rightarrow \frac{du}{dx} = -\sin x\ \frac{dv}{dx} = \cos x e^{\sin x}$

Product rule $\Rightarrow f'(x) = -\sin x e^{\sin x} + \cos^2 x e^{\sin x}$
$= e^{\sin x}(\cos^2 x - \sin x)$

c) $f''(x) = 0 \Rightarrow e^{\sin x}(\cos^2 x - \sin x) = 0$

$\Rightarrow \cos^2 x - \sin x = 0 \Rightarrow 1 - \sin^2 x - \sin x = 0$

$\sin^2 x + \sin x - 1 = 0 \Rightarrow \sin x = \frac{-1 \pm \sqrt{1+4}}{2}$

$\Rightarrow \sin x = \frac{-1 \pm \sqrt{5}}{2}$ $\quad (e^{\sin x} \neq 0)$

Since $\frac{-1-\sqrt{5}}{2} < -1$ then $\sin x = \frac{-1+\sqrt{5}}{2}$

Mechanics

1 Draw a diagram to help you!!

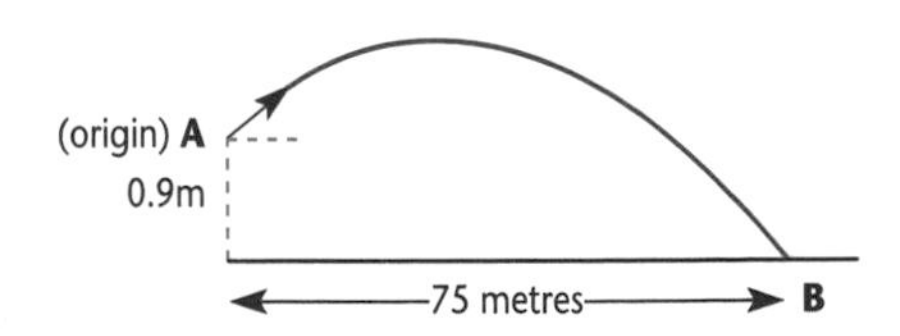

a) AB(↑) : $u_v = ??$, $s = -0.9$, $a = -9.8$, $t = 3$

$s = ut + \frac{1}{2}at^2 \Rightarrow -0.9 = 3u_v - 4.9(9)$

$\Rightarrow 43.2 = 3u_v \Rightarrow u_v = 14.4\ \text{ms}^{-1}$

AB (→): $s = 75$, $u_h = ??$, $t = 3$.

Hence $s = ut \Rightarrow 75 = 3u_h \Rightarrow u_h = 25\ \text{ms}^{-1}$

Use your Knowledge Answers

b)

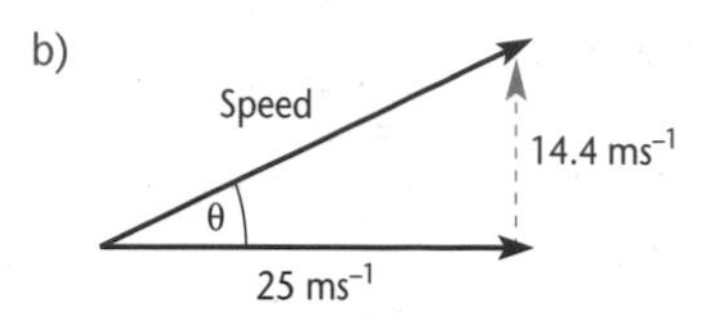

speed = magnitude of initial velocity

$= \sqrt{(25)^2 + (14.4)^2} = 28.851\ldots = 28.9\ \text{ms}^{-1}$ (3 sf)

direction $= \theta = \tan^{-1}\left(\frac{14.4}{25}\right) = 29.942\ldots = 29.9°$ (3 sf)

c) Resolve for A (↑): $u_v = 14.4$, $s = 6 - 0.9 = 5.1$, $a = -9.8$,

$t = ??: \Rightarrow s = ut + \frac{1}{2}at^2$

$5.1 = 14.4t - 4.9t^2 : 4.9t^2 - 14.4t + 5.1 = 0$

Formula: $t = \dfrac{14.4 \pm \sqrt{(14.4)^2 - (4 \times 4.9 \times 5.1)}}{2 \times 4.9}$

$t = \dfrac{14.4 \pm \sqrt{107.4}}{9.8} = 0.412\ldots, 2.527\ldots$

$\Rightarrow$ Duration $= 2.527 - 0.412 = 2.1$ s (1d.p.)

2 a) 5260.8 W

b) 14.6 ms^{-1}

3 a) 1.6 cm from AD; 2.167cm from BE

b) 17°

Statistics

1 a) B(600, 0.75)

b) Use N(450, 112.5)

i) P(X > 429.5)

$Z = \dfrac{429.5 - 450}{\sqrt{112.5}} = -1.93$

P(X > 429.5) = 0.9732

ii) P(439.5 < X < 460.5)

$Z = \dfrac{439.5 - 450}{\sqrt{112.5}} = -0.99$

$Z = \dfrac{460.5 - 450}{\sqrt{112.5}} = 0.99$

P(439.5 < X < 460.5) = 0.8389 − (1 − 0.8389)

= 0.6778

2 X = number of breakdowns

a) $X \sim \text{Poi}(3.5)$, per week

b) $P(X > 5) = 1 - P(X \leq 5) = 1 - 0.8576 = 0.1424 = 0.142$ (3sf)

c) $X \sim \text{Poi}(0.5)$, per day $\Rightarrow P(X = 0) = 0.6065 = 0.607$ (3sf)

d) For 1 year, $X \sim \text{Poi}(182) \to X \sim N(182, 182)$.
Hence $P(X \geq 200) \cong P(X > 199.5)$ (by c.c.)

z-value: $\dfrac{199.5 - 182}{\sqrt{182}} = 1.297\ldots : P(Z > 1.30) = 1 - \Phi(1.30)$
$= 0.0968$ (3sf) (0.0973)

3 a) X = time taken to complete exercise $\Rightarrow X \sim N(48, 5.2^2)$.

Hence: P(X < 52)

z-value: $\dfrac{52 - 48}{5.2} = 0.769\ldots$

$\Rightarrow P(z < 0.77) = \Phi(0.77) = 0.7794 = 0.779$ (3sf) (0.779)

b) Drawing a normal curve yields the equation:

$\dfrac{t - 48}{5.2} = -2.3263 : t = 48 - (2.3263 \times 5.2)$

$t = 35.9$ mins (3sf)

c) Y = number who win GM prize
$\Rightarrow Y \sim \text{Bin}(310, 0.01) \times Y \sim \text{Poi}(3.1)$

$P(Y < 3) = P(Y = 0) + P(Y = 1) + P(Y = 2)$

$= \dfrac{e^{-3.1}(3.1)^0}{0!} + \dfrac{e^{-3.1}(3.1)^1}{1!} + \dfrac{e^{-3.1}(3.1)^2}{2!}$

$= 0.04505\ldots + 0.13965\ldots + 0.21646\ldots = 0.40116\ldots$
$= 0.401$ (3 sf)

d) E = number passing physical exercise

$\Rightarrow E \sim \text{Bin}(310, 0.779) \to E \sim N(241.49, 53.369\ldots)$

Hence: $P(E \geq 250) \cong P(E > 249.5)$ (by c.c.)

z-value: $\dfrac{249.5 - 241.49}{\sqrt{53.369\ldots}} = 1.096\ldots$

$P(z > 1.10) = 1 - \Phi(1.10) = 1 - 0.8643$

$= 0.1357 = 0.136$ (3sf) (0.137)